Nonlinear Differential Equations and Dynamical Systems

Nonlinear Differential Equations and Dynamical Systems

Theory and Applications

Editors

Feliz Manuel Minhós
João Fialho

MDPI • Basel • Beijing • Wuhan • Barcelona • Belgrade • Manchester • Tokyo • Cluj • Tianjin

Editors
Feliz Manuel Minhós
University of Évora
Portugal

João Fialho
British University Vietnam
Vietnam

Editorial Office
MDPI
St. Alban-Anlage 66
4052 Basel, Switzerland

This is a reprint of articles from the Special Issue published online in the open access journal *Axioms* (ISSN 2075-1680) (available at: https://www.mdpi.com/journal/axioms/special_issues/nonlinear_differential_equations).

For citation purposes, cite each article independently as indicated on the article page online and as indicated below:

LastName, A.A.; LastName, B.B.; LastName, C.C. Article Title. *Journal Name* **Year**, *Volume Number*, Page Range.

ISBN 978-3-0365-0710-1 (Hbk)
ISBN 978-3-0365-0711-8 (PDF)

© 2021 by the authors. Articles in this book are Open Access and distributed under the Creative Commons Attribution (CC BY) license, which allows users to download, copy and build upon published articles, as long as the author and publisher are properly credited, which ensures maximum dissemination and a wider impact of our publications.

The book as a whole is distributed by MDPI under the terms and conditions of the Creative Commons license CC BY-NC-ND.

Contents

About the Editors vii

Preface to "Nonlinear Differential Equations and Dynamical Systems" ix

Nipon Waiyaworn, Kamsing Nonlaopon and Somsak Orankitjaroen
Finite Series of Distributional Solutions for Certain Linear Differential Equations
Reprinted from: *Axioms* **2020**, *9*, 116, doi:10.3390/axioms9040116 1

Ravi P. Agarwal, Petio S. Kelevedjiev and Todor Z. Todorov
On the Solvability of Nonlinear Third-Order Two-Point Boundary Value Problems
Reprinted from: *Axioms* **2020**, *9*, 62, doi:10.3390/axioms9020062 13

Abdukomil Risbekovich Khashimov and Dana Smetanová
On the Uniqueness Classes of Solutions of Boundary Value Problems for Third-Order Equations of the Pseudo-Elliptic Type
Reprinted from: *Axioms* **2020**, *9*, 80, doi:10.3390/axioms9030080 23

Omar Bazighifan, Feliz Minhos and Osama Moaaz
Sufficient Conditions for Oscillation of Fourth-Order Neutral Differential Equations with Distributed Deviating Arguments
Reprinted from: *Axioms* **2020**, *9*, 39, doi:10.3390/axioms9020039 31

Omar Bazighifan, Rami Ahmad El-Nabulsi and Osama Moaaz
Asymptotic Properties of Neutral Differential Equations with Variable Coefficients
Reprinted from: *Axioms* **2020**, *9*, 96, doi:10.3390/axioms9030096 43

Choukri Derbazi, Zidane Baitiche, Mouffak Benchohra and Alberto Cabada
Initial Value Problem For NonlinearFractional Differential Equations With ψ-Caputo Derivative via Monotone Iterative Technique
Reprinted from: *Axioms* **2020**, *9*, 57, doi:10.3390/axioms9020057 55

Tursun K. Yuldashev, Bakhtiyor J. Kadirkulov
Boundary Value Problem for Weak NonlinearPartial Differential Equations of Mixed Type with Fractional Hilfer Operator
Reprinted from: *Axioms* **2020**, *9*, 68, doi:10.3390/axioms9020068 69

Hasan S. Panigoro, Agus Suryanto, Wuryansari Muharini Kusumawinahyu and Isnani Darti
A Rosenzweig–MacArthur Model with Continuous Threshold Harvesting in Predator Involving Fractional Derivatives with Power Law and Mittag–Leffler Kernel
Reprinted from: *Axioms* **2020**, *9*, 122, doi:10.3390/axioms9040122 89

Tursun K. Yuldashev and Erkinjon T. Karimov
Inverse Problem for a Mixed Type Integro-Differential Equation with Fractional Order Caputo Operators and Spectral Parameters
Reprinted from: *Axioms* **2020**, *9*, 121, doi:10.3390/axioms9040121 111

Faïçal Ndaïrou and Delfim F. M. Torres
Distributed-Order Non-Local Optimal Control
Reprinted from: *Axioms* **2020**, *9*, 124, doi:10.3390/axioms9040124 135

About the Editors

Feliz Manuel Minhós has a PhD and Habilitation Degree, both in Mathematics, is the Coordinador of the Research Center on Mathematics and Applications, Director of the PhD Program in Mathematics of the University of Évora, Portugal, has published a large number of papers, books, monographs, proceedings, etc., and is member of the Editorial Board of several international journals. The main research interests are in Differential Equations, Boundary Value Problems with variational and topological methods, Fixed Point theory, Impulsive problems, functional differential equations, nonlocal problems, degree theory, lower and upper solutions and Green´s functions, among others.

João Fialho is a PhD in Mathematics of University of Évora, Portugal. He is currently Senior Lecturer and MBA and Graduate Programmes Director at the British University of Vietnam. He has diverse practical working experience and teaching experience, in the areas of actuarial sciences, pricing, finance and data analysis, accumulated at the service of several multinational companies. In the last 8 years, he has accumulated experience in Higher Education, having already taught in North America, Europe, Middle East and Asia, accumulating positions of Department Director and research coordination. Prof. Fialho is an active researcher in the field of differential equations, mathematical modelling, big data and data analytics, with proven experience by a wide range of international publications, books and various communications at international conferences, in the area of mathematics and applications.

Preface to "Nonlinear Differential Equations and Dynamical Systems"

Nonlinear differential equations, dynamical systems, and related topics are particularly trendy topics currently, as they have had wide and significant applications in many fields of Physics, Chemistry, Engineering, Biology, or even Economics, in general, and Mathematics in particular.

They can be approached using several different methods and techniques. As examples, we can refer to variational and topological methods, fractional derivatives, fixed point theory, initial and boundary value problems, qualitative theory, stability theory, existence and control of chaos, the existence of attractors and periodic orbits, among others.

This Special Issue contains original results and recent developments in some of the above fields, such as fractional differential and integral equations and applications, non-local optimal control, inverse, and higher-order nonlinear boundary value problems, distributional solutions in the form of a finite series of the Dirac delta function and its derivatives, asymptotic properties oscillatory theory for neutral nonlinear differential equations, the existence of extremal solutions via monotone iterative techniques, and predator–prey interaction via fractional-order models, among others.

These recent results, and the diversity of methods and themes, involving new trends in several areas of mathematical research, allow the reader a glance at the related state-of-the-art, and may provide interested researchers with ideas and techniques that lead to new research and new results.

Feliz Manuel Minhós, João Fialho
Editors

Article

Finite Series of Distributional Solutions for Certain Linear Differential Equations

Nipon Waiyaworn [1], Kamsing Nonlaopon [1,*] and Somsak Orankitjaroen [2]

[1] Department of Mathematics, Khon Kaen University, Khon Kaen 40002, Thailand; nipon_waiyaworn@kkumail.com
[2] Department of Mathematics, Faculty of Science, Mahidol University, Bangkok 10400, Thailand; somsak.ora@mahidol.ac.th
* Correspondence: nkamsi@kku.ac.th; Tel.: +66-8-6642-1582

Received: 31 August 2020; Accepted: 2 October 2020; Published: 13 October 2020

Abstract: In this paper, we present the distributional solutions of the modified spherical Bessel differential equations $t^2 y''(t) + 2t y'(t) - [t^2 + \nu(\nu+1)] y(t) = 0$ and the linear differential equations of the forms $t^2 y''(t) + 3t y'(t) - (t^2 + \nu^2 - 1) y(t) = 0$, where $\nu \in \mathbb{N} \cup \{0\}$ and $t \in \mathbb{R}$. We find that the distributional solutions, in the form of a finite series of the Dirac delta function and its derivatives, depend on the values of ν. The results of several examples are also presented.

Keywords: Dirac delta function; distributional solution; Laplace transform; power series solution

1. Introduction

It is well known that the linear differential equation of the form

$$\sum_{n=0}^{m} a_n(t) y^{(n)}(t) = 0, \quad a_m(t) \neq 0, \tag{1}$$

where $a_n(t)$ is an infinitely smooth coefficient for each n, and has no distributional solutions other than the classical ones. However, if the leading coefficient $a_m(t)$ has a zero, the classical solution of (1) may cease to exist in a neighborhood of that zero. In that case, (1) may have a distributional solution. It was not until 1982 that Wiener [1] proposed necessary and sufficient conditions for the existence of an Nth-order distributional solution to the differential equation (1). The Nth-order distributional solution that Wiener proposed is a finite sum of Dirac delta function and its derivatives:

$$y(t) = \sum_{n=0}^{N} b_n \delta^{(n)}(t), \quad b_N \neq 0. \tag{2}$$

It can be easily verified by (10) that $\delta(t)$ is a zero order distributional solution of the equation

$$t y''(t) + 2 y'(t) + t y(t) = 0;$$

the Bessel equation

$$t^2 y''(t) + t y'(t) + (t^2 - 1) y(t) = 0;$$

the confluent hypergeometric equation

$$t y''(t) + (2 - t) y'(t) - y(t) = 0;$$

and the second order Cauchy–Euler equation

$$t^2 y''(t) + 3t y'(t) + y(t) = 0.$$

The distributional solutions with higher order of Cauchy–Euler equations were studied by many researchers; see [2–8] for more details.

The infinite order distributional solution of the form

$$y(t) = \sum_{n=0}^{\infty} b_n \delta^{(n)}(t) \tag{3}$$

to various differential equations in a normal form with singular coefficients was studied by many researchers [9–13]. Furthermore, a brief introduction to these concepts is presented by Kanwal [14].

In 1984, Cooke and Wiener [15] presented the existence theorems for distributional and analytic solutions of functional differential equations. In 1987, Littlejohn and Kanwal [16] studied the distributional solutions of the hypergeometric differential equation, whose solutions are in the form of (3). In 1990, Wiener and Cooke [17] presented the necessary and sufficient conditions for the simultaneous existence of solutions to linear ordinary differential equations in the forms of rational functions and (2).

As mentioned in abstract, we propose the distributional solutions of the modified spherical Bessel differential equations

$$t^2 y''(t) + 2t y'(t) - [t^2 + \nu(\nu+1)] y(t) = 0$$

and the linear differential equations of the forms

$$t^2 y''(t) + 3t y'(t) - (t^2 + \nu^2 - 1) y(t) = 0,$$

where $\nu \in \mathbb{N} \cup \{0\}$ and $t \in \mathbb{R}$. The modified spherical Bessel differential equation is just the spherical Bessel equation with a negative separation constant. The spherical Bessel equation occurs when dealing with the Helmholtz equation in spherical coordinates of various problems in physics such as a scattering problem [18].

We use the simple method, consisting of Laplace transforms of right-sided distributions and power series solution, for searching the distributional solutions of these equations. We find that the solutions are in the forms of finite linear combinations of the Dirac delta function and its derivatives depending on the values of ν.

2. Preliminaries

In this section, we introduce the basic knowledge and concepts, which are essential for this work.

Definition 1. *Let $\mathcal{D}$ be the space consisting of all real-valued functions $\varphi(t)$ with continuous derivatives of all orders and compact support. The support of $\varphi(t)$ is the closure of the set of all elements $t \in \mathbb{R}$ such that $\varphi(t) \neq 0$. Then $\varphi(t)$ is called a test function.*

Definition 2. *A distribution T is a continuous linear functional on the space $\mathcal{D}$. The space of all such distributions is denoted by $\mathcal{D}'$.*

For every $T \in \mathcal{D}'$ and $\varphi(t) \in \mathcal{D}$, the value that T acts on $\varphi(t)$ is denoted by $\langle T, \varphi(t) \rangle$. Note that $\langle T, \varphi(t) \rangle \in \mathbb{R}$.

Example 1.

(i) The locally integrable function $f(t)$ is a distribution generated by the locally integrable function $f(t)$. Then we define $\langle f(t), \varphi(t) \rangle = \int_\Omega f(t) \varphi(t) dt$, where Ω is the support of $\varphi(t)$ and $\varphi(t) \in \mathcal{D}$.

(ii) The Dirac delta function is a distribution defined by $\langle \delta(t), \varphi(t) \rangle = \varphi(0)$ and the support of $\delta(t)$ is $\{0\}$.

A distribution T generated by a locally integrable function is called a regular distribution; otherwise, it is called a singular distribution.

Definition 3. *The kth-order derivative of a distribution T, denoted by $T^{(k)}$, is defined by $\left\langle T^{(k)}, \varphi(t) \right\rangle = (-1)^k \left\langle T, \varphi^{(k)}(t) \right\rangle$ for all $\varphi(t) \in \mathcal{D}$.*

Example 2.

(i) $\langle \delta'(t), \varphi(t) \rangle = -\langle \delta(t), \varphi'(t) \rangle = -\varphi'(0);$

(ii) $\left\langle \delta^{(k)}(t), \varphi(t) \right\rangle = (-1)^k \left\langle \delta(t), \varphi^{(k)}(t) \right\rangle = (-1)^k \varphi^{(k)}(0).$

Definition 4. *Let $\alpha(t)$ be an infinitely differentiable function. We define the product of $\alpha(t)$ with any distribution T in $\mathcal{D}'$ by $\langle \alpha(t)T, \varphi(t) \rangle = \langle T, \alpha(t)\varphi(t) \rangle$ for all $\varphi(t) \in \mathcal{D}$.*

Definition 5. *If $y(t)$ is a singular distribution and satisfies the equation*

$$\sum_{m=0}^{n} a_m(t) y^{(n)}(t) = f(t), \qquad (4)$$

where $a_m(t)$ is an infinitely differentiable function and $f(t)$ is an arbitrary known distribution, in the sense of distribution, and is called a distributional solution of (4).

Definition 6. *Let $M \in \mathbb{R}$ and $f(t)$ be a locally integrable function satisfying the following conditions:*

(i) $f(t) = 0$ for all $t < M$;

(ii) *There exists a real number c such that $e^{-ct} f(t)$ is absolutely integrable over $\mathbb{R}$.*

The Laplace transform of $f(t)$ is defined by

$$F(s) = \mathcal{L}\{f(t)\} = \int_M^\infty f(t) e^{-st} dt, \qquad (5)$$

where s is a complex variable.

It is well known that if $f(t)$ is continuous, then $F(s)$ is an analytic function on the half-plane $\Re(s) > \sigma_a$, where σ_a is an abscissa of absolute convergence for $\mathcal{L}\{f(t)\}$.

Recall that the Laplace transform $G(s)$ of a locally integrable function $g(t)$ satisfying the conditions of definition 6, that is,

$$G(s) = \mathcal{L}\{g(t)\} = \int_M^\infty g(t) e^{-st} dt, \qquad (6)$$

where $\Re(s) > \sigma_a$, can be written in the form $G(s) = \left\langle g(t), e^{-st} \right\rangle$.

Definition 7. *Let S be the space of test functions of rapid decay containing the complex-valued functions $\phi(t)$ having the following properties:*

(i) $\phi(t)$ is infinitely differentiable—i.e., $\phi(t) \in C^\infty(\mathbb{R})$;

(ii) $\phi(t)$, as well as its derivatives of all orders, vanish at infinity faster than the reciprocal of any polynomial which is expressed by the inequality

$$|t^p \phi^{(k)}(t)| < C_{pk},$$

where C_{pk} is a constant depending on $p, k,$ and $\phi(t)$. Then $\phi(t)$ is called a test function in the space S.

Definition 8. *A distribution of slow growth or tempered distribution T is a continuous linear functional over the space S of test function of rapid decay and contains the complex-valued functions—i.e., there is assigned a complex number $\langle T, \phi(t) \rangle$ with properties:*

(i) $\langle T, c_1\phi_1(t) + c_2\phi_2(t) \rangle = c_1 \langle T, \phi_1(t) \rangle + c_2 \langle T, \phi_2(t) \rangle$ for $\phi_1(t), \phi_2(t) \in S$ and constants c_1, c_2;
(ii) $\lim_{m \to \infty} \langle T, \phi_m(t) \rangle = 0$ for every null sequence $\{\phi_m(t)\} \in S$.

We shall let S' denote the set of all distributions of slow growth.

Definition 9. *Let $f(t)$ be a distribution satisfying the following properties:*

(i) $f(t)$ is a right-sided distribution, that is, $f(t) \in \mathcal{D}'_R$.
(ii) There exists a real number c such that $e^{-ct} f(t)$ is a tempered distribution.

The Laplace transform of a right-sided distribution $f(t)$ satisfying (ii) is defined by

$$F(s) = \mathcal{L}\{f(t)\} = \left\langle e^{-ct} f(t), X(t) e^{-(s-c)t} \right\rangle, \tag{7}$$

where $X(t)$ is an infinitely differentiable function with support bounded on the left, which equals 1 over a neighbourhood of the support of $f(t)$.

For $\Re(s) > c$, the function $X(t) e^{-(s-c)t}$ is a testing function in the space S and $e^{-ct} f(t)$ is in the space S'. Equation (7) can be reduced to

$$F(s) = \mathcal{L}\{f(t)\} = \left\langle f(t), e^{-st} \right\rangle. \tag{8}$$

Now $F(s)$ is a function of s defined over the right half-plane $\Re(s) > c$. Zemanian [19] proved that $F(s)$ is an analytic function in the region of convergence $\Re(s) > \sigma_1$, where σ_1 is the abscissa of convergence and $e^{-ct} f(t) \in S'$ for some real number $c > \sigma_1$.

Example 3. *Let $\delta(t)$ be the Dirac delta function, $H(t)$ be the Heaviside function, and $f(t)$ be a Laplace-transformable distribution in $\mathcal{D}'_R$. If k is a positive integer, then the following holds:*

(i) $\mathcal{L}\{(t^{k-1} H(t))/(k-1)!\} = 1/s^k, \quad \Re(s) > 0$;
(ii) $\mathcal{L}\{\delta(t)\} = 1, \quad -\infty < \Re(s) < \infty$;
(iii) $\mathcal{L}\left\{\delta^{(k)}(t)\right\} = s^k, \quad -\infty < \Re(s) < \infty$;
(iv) $\mathcal{L}\left\{t^k f(t)\right\} = (-1)^k F^{(k)}(s), \quad \Re(s) > \sigma_1$;
(v) $\mathcal{L}\left\{f^{(k)}(t)\right\} = s^k F(s), \quad \Re(s) > \sigma_1$.

The proof of following lemma 1 is given in [14].

Lemma 1. *Let $\psi(t)$ be an infinitely differentiable function. Then*

$$\psi(t)\delta^{(m)}(t) = (-1)^m \psi^{(m)}(0)\delta(t) + (-1)^{m-1} m \psi^{(m-1)}(0)\delta'(t)$$
$$+ (-1)^{m-2} \frac{m(m-1)}{2!} \psi^{(m-1)}(0)\delta''(t) + \cdots + \psi(0)\delta^{(m)}(t). \tag{9}$$

A useful formula that follows from (9), for any monomial $\psi(t) = t^n$, is that

$$t^n \delta^{(m)}(t) = \begin{cases} 0, & \text{if } m < n; \\ (-1)^n \frac{m!}{(m-n)!} \delta^{(m-n)}(t), & \text{if } m \geq n. \end{cases} \tag{10}$$

3. Main Results

In this section, we will state our main results and give their proofs.

Theorem 1. *Consider the differential equation of the form*

$$t^2 y''(t) + 2t y'(t) - [t^2 + \nu(\nu+1)] y(t) = 0, \tag{11}$$

where $\nu \in \mathbb{N} \cup \{0\}$ and $t \in \mathbb{R}$. The distributional solutions of (11) are given by

$$y(t) = P_\nu(D) \delta(t), \tag{12}$$

where

$$P_\nu(D) = \frac{1}{2^\nu} \sum_{k=0}^{\lfloor \nu/2 \rfloor} (-1)^k \frac{(2\nu - 2k)!}{k!(\nu-k)!(\nu-2k)!} D^{\nu-2k},$$

is a Legendre polynomial of distributional derivative operator $D = d/dt$.

Proof. Applying the Laplace transform to both sides of (11) with $\mathcal{L}\{y(t)\} = F(s)$, and using Example 3(iv), (v), we obtain

$$(1 - s^2) F''(s) - 2s F'(s) + \nu(\nu+1) F(s) = 0. \tag{13}$$

Suppose that a solution of (13) is of the form $F(s) = \sum_{n=0}^{\infty} a_n s^n$. Differentiating term by term, we obtain

$$F'(s) = \sum_{n=1}^{\infty} n a_n s^{n-1}$$

and

$$F''(s) = \sum_{n=2}^{\infty} n(n-1) a_n s^{n-2}.$$

Substituting these terms into (13), we have

$$[2 a_2 + \nu(\nu+1) a_0] + [(3 \cdot 2) a_3 - (2 - \nu(\nu+1)) a_1] s$$
$$+ \sum_{n=2}^{\infty} [(n+2)(n+1) a_{n+2} - n(n-1) a_n - 2 n a_n + \nu(\nu+1) a_n] s^n = 0.$$

Since $s^n \neq 0$, it follows that

$$2 a_2 + \nu(\nu+1) a_0 = 0, \quad (3 \cdot 2) a_3 - (2 - \nu(\nu+1)) a_1 = 0,$$
$$(n+2)(n+1) a_{n+2} - n(n-1) a_n - 2 n a_n + \nu(\nu+1) a_n = 0, \quad n \geq 2,$$

which leads to a recurrence relation

$$a_{n+2} = -\frac{(\nu-n)(\nu+n+1)}{(n+1)(n+2)} a_n. \tag{14}$$

Thus, we obtain

$$a_2 = -\frac{\nu(\nu+1)}{2!}a_0$$
$$a_4 = (-1)^2 \frac{\nu(\nu-2)(\nu+1)(\nu+3)}{4!}a_0$$
$$\vdots$$
$$a_{2n} = (-1)^n \frac{\nu(\nu-2)\cdots(\nu-2n+2)(\nu+1)(\nu+3)\cdots(\nu+2n-1)}{(2n)!}a_0.$$

Similarly,

$$a_3 = -\frac{(\nu-1)(\nu+2)}{2\cdot 3}a_1$$
$$a_5 = (-1)^2 \frac{(\nu-1)(\nu-3)(\nu+2)(\nu+4)}{5!}a_1$$
$$\vdots$$
$$a_{2n+1} = (-1)^n \frac{(\nu-1)(\nu-3)\cdots(\nu-2n+1)(\nu+2)(\nu+4)\cdots(\nu+2n)}{(2n+1)!}a_1.$$

Letting $a_0 = a_1 = 1$, we get the two solutions of (13) in the forms

$$F_e(s) = 1 + \sum_{n=1}^{\infty}(-1)^n \frac{\nu(\nu-2)\cdots(\nu-2n+2)(\nu+1)(\nu+3)\cdots(\nu+2n-1)}{(2n)!}s^{2n}$$

and

$$F_o(s) = s + \sum_{n=1}^{\infty}(-1)^n \frac{(\nu-1)(\nu-3)\cdots(\nu-2n+1)(\nu+2)(\nu+4)\cdots(\nu+2n)}{(2n+1)!}s^{2n+1}.$$

If ν is even, letting $\nu = 2m$, $m \in \mathbb{N} \cup \{0\}$, we note that

$$\nu(\nu-2)\cdots(\nu-2n+2) = 2m(2m-2)\cdots(2m-2n+2) = \begin{cases} 0, & m=0; \\ \dfrac{2^n m!}{(m-n)!}, & m>0,\ n \leq m; \\ 0, & n > m > 0, \end{cases}$$

and

$$(\nu+1)(\nu+3)\cdots(\nu+2n-1) = (2m+1)(2m+3)\cdots(2m+2n-1) = \frac{(2m+2n)!m!}{2^n(2m)!(m+n)!}.$$

Then, in this case, $F_e(s)$ only becomes the finite series of the form

$$F_e(s) = \frac{(m!)^2}{(2m)!}\sum_{k=0}^{m}(-1)^k \frac{(2m+2k)!s^{2k}}{(m-k)!(m+k)!(2k)!}. \tag{15}$$

If ν is odd, letting $\nu = 2m+1$, $m \in \mathbb{N} \cup \{0\}$, we note that

$$(\nu-1)(\nu-3)\cdots(\nu-2n+1) = 2m(2m-2)\cdots(2m-2n+2) = \begin{cases} 0, & m=0; \\ \dfrac{2^n m!}{(m-n)!}, & m>0,\ n \leq m; \\ 0, & n > m > 0, \end{cases}$$

and

$$(\nu+2)(\nu+4)\cdots(\nu+2n) = (2m+3)(2m+5)\cdots(2m+2n+1) = \frac{(2m+2n+1)!m!}{2^n(2m+1)!(m+n)!}.$$

Then, in this case, $F_o(s)$ only becomes the finite series of the form

$$F_o(s) = \frac{(m!)^2}{(2m+1)!} \sum_{k=0}^{m} (-1)^k \frac{(2m+2k+1)!s^{2k+1}}{(m-k)!(m+k)!(2k+1)!}. \tag{16}$$

For $\nu = 0, 1, 2, \ldots$, we have $F_\nu(s)$, as follows:

$$F_0(s) = 1 = P_0(s),$$
$$F_1(s) = s = P_1(s),$$
$$F_2(s) = 1 - 3s^2 = -2P_2(s),$$
$$F_3(s) = s - (5/3)s^3 = -(2/3)P_3(s),$$
$$F_4(s) = 1 - 10s^2 + (35/3)s^4 = (8/3)P_4(s),$$
$$F_5(s) = s - (14/3)s^3 + (21/5)s^5 = (8/15)P_5(s),$$
$$F_6(s) = 1 - 21s^2 + 63s^4 - (231/5)s^6 = -(16/5)P_6(s),$$
$$F_7(s) = s - 9s^3 + (99/5)s^5 - (429/35)s^7 = -(16/35)P_7(s),$$

and so on, where $P_n(s)$ is the Legendre polynomial of s for $n = 0, 1, 2, \ldots$. Since (13) is linear, P_ν is also its solution for all non-negative integer ν. Taking the inverse Laplace transform to $P_\nu(s)$, and using Example 3(ii),(iii), we obtain the solutions of (11),

$$y(t) = \frac{1}{2^\nu} \sum_{k=0}^{\lfloor \nu/2 \rfloor} (-1)^k \frac{(2\nu-2k)!}{k!(\nu-k)!(\nu-2k)!} D^{\nu-2k}\delta(t), \quad (D \equiv \frac{d}{dt} \text{ distribution derivative}) \tag{17}$$
$$= P_\nu(D)\delta(t),$$

which are the distributional solutions of the form (12). □

Example 4. Letting $\nu = 1$, (11) becomes

$$t^2 y''(t) + 2ty'(t) - (t^2+2)y(t) = 0. \tag{18}$$

From Theorem 1, (18) has a solution

$$y(t) = \delta'(t). \tag{19}$$

Letting $\nu = 4$, (11) becomes

$$t^2 y''(t) + 2ty'(t) - (t^2+20)y(t) = 0. \tag{20}$$

From Theorem 1, (20) has a solution

$$y(t) = \frac{35}{8}\delta^{(4)}(t) - \frac{15}{4}\delta''(t) + \frac{3}{8}\delta(t). \tag{21}$$

By applying (10), it is not difficult to verify that (19) and (21) satisfy (18) and (20), respectively.

Theorem 2. Consider the equation of the form

$$t^2 y''(t) + 3ty'(t) - (t^2 + \nu^2 - 1)y(t) = 0, \tag{22}$$

where $v \in \mathbb{N} \cup \{0\}$ and $t \in \mathbb{R}$. The distributional solutions of (22) are given by

$$y(t) = T_v(D)\delta(t), \tag{23}$$

where

$$T_v(D) = \frac{v}{2} \sum_{k=0}^{\lfloor v/2 \rfloor} (-1)^k \frac{2^{v-2k}(n-k-1)!}{k!(v-2k)!} D^{v-2k},$$

is a Chebyshev polynomial of the first kind of distributional derivative operator $D = d/dt$.

Proof. Applying the Laplace transform $\mathcal{L}\{y(t)\} = F(s)$ to (22) and using Example 3(iv), (v), we obtain

$$(1-s^2)F''(s) - sF'(s) + v^2 F(s) = 0. \tag{24}$$

Suppose that a solution of (24) is of the form $F(s) = \sum_{n=0}^{\infty} a_n s^n$. Differentiating $F(s)$ term by term, we obtain

$$F'(s) = \sum_{n=1}^{\infty} n a_n s^{n-1}$$

and

$$F''(s) = \sum_{n=2}^{\infty} n(n-1) a_n s^{n-2}.$$

Substituting these terms into (24), we get

$$2a_2 + v^2 a_0 + \left(6a_3 - a_1 + v^2 a_1\right) s + \sum_{n=2}^{\infty} \left\{(n+2)(n+1)a_{n+2} - [n(n-1) + n - v^2]a_n\right\} s^n = 0.$$

Since $s^n \neq 0$, it follows that

$$2a_2 + v^2 a_0 = 0,$$
$$6a_3 - a_1 + v^2 a_1 = 0,$$

and for $n = 2, 3, \ldots$,

$$(n+2)(n+1)a_{n+2} - (n^2 - v^2)a_n = 0.$$

Hence,

$$a_2 = -\frac{v^2}{2} a_0,$$
$$a_3 = \frac{1-v^2}{3!} a_1,$$

and for $n = 2, 3, \ldots$,

$$a_{n+2} = \frac{n^2 - v^2}{(n+2)(n+1)} a_n. \tag{25}$$

Thus, we obtain

$$a_2 = -\frac{v^2}{2} a_0,$$
$$a_4 = \frac{(2^2 - v^2)(-v^2)}{4!} a_0,$$

and so on. Similarly,

$$a_3 = \frac{1-\nu^2}{3!}a_1,$$
$$a_5 = \frac{(3^2-\nu^2)(1-\nu^2)}{5!}a_1,$$

and so on. A pattern clearly emerges:

$$a_{2n} = \frac{[(2n-2)^2-\nu^2][(2n-4)^2-\nu^2]\cdots(2^2-\nu^2)(-\nu^2)}{(2n)!}a_0$$

and

$$a_{2n+1} = \frac{[(2n-1)^2-\nu^2][(2n-3)^2-\nu^2]\cdots(3^2-\nu^2)(1-\nu^2)}{(2n+1)!}a_1.$$

Hence, we get two solutions of (24) in the forms

$$F_e(s) = a_0 + \sum_{n=1}^{\infty} \frac{[(2n-2)^2-\nu^2][(2n-4)^2-\nu^2]\cdots(2^2-\nu^2)(-\nu^2)}{(2n)!}a_0 s^{2n}$$

and

$$F_o(s) = a_1 s + \sum_{n=1}^{\infty} \frac{[(2n-1)^2-\nu^2][(2n-3)^2-\nu^2]\cdots(3^2-\nu^2)(1-\nu^2)}{(2n+1)!}a_1 s^{2n+1}.$$

If ν is even, letting $\nu = 2m$, $m \in \mathbb{N} \cup \{0\}$, then $a_{2m+2} = 0$, so that $a_{2n+2} = 0$ for $n \geq m$. Hence,

$$F_e(s) = 1 + \sum_{n=1}^{m} \frac{[(2n-2)^2-(2m)^2][(2n-4)^2-(2m)^2]\cdots[2^2-(2m)^2][-(2m)^2]}{(2n)!}s^{2n},$$

equivalently to

$$F_e(s) = 1 + \sum_{n=1}^{m} \frac{\prod_{k=1}^{n} 4(n-k+m)(n-k-m)}{(2n)!}s^{2n}.$$

If ν is odd, letting $\nu = 2m+1$, $m \in \mathbb{N} \cup \{0\}$, then $a_{2m+3} = 0$, so that $a_{2n+3} = 0$ for $n \geq m$. Hence,

$$F_o(s) = s + \sum_{n=1}^{m} \frac{[(2n-1)^2-(2m+1)^2][(2n-3)^2-(2m+1)^2]\cdots[3^2-(2m+1)^2)][1-(2m+1)^2}{(2n+1)!}s^{2n+1},$$

equivalently to

$$F_o(s) = s + \sum_{n=1}^{m} \frac{\prod_{k=1}^{n} 4(n-k+m+1)(n-k-m)}{(2n+1)!}s^{2n+1}.$$

For $\nu = 0, 1, 2, \ldots$, we have $F_\nu(s)$, as follows:

$$F_0(s) = 1 = T_0(s),$$
$$F_1(s) = s = T_1(s),$$
$$F_2(s) = 1 - 2s^2 = -T_2(s),$$
$$F_3(s) = s - (4/3)s^3 = -(1/3)T_3(s),$$
$$F_4(s) = 1 - 8s^2 + 8s^4 = T_4(s),$$
$$F_5(s) = s - 4s^3 + (16/5)s^5 = (1/5)T_5(s),$$
$$F_6(s) = 1 - 18s^2 + 48s^4 - 32s^6 = -T_6(s),$$
$$F_7(s) = s - 8s^3 + 16s^5 - (64/7)s^7 = -(1/7)T_7(s),$$

and so on, where $T_n(s)$ is the Chebyshev polynomial of the first kind of s for $n = 0, 1, 2, \ldots$. Since (24) is linear, $T_n(s)$ is also its solution. Taking the inverse Laplace transform to $T_n(s)$, and using Example 3(ii),(iii), we obtain the solutions of (22),

$$y_n(t) = \frac{\nu}{2} \sum_{k=0}^{\lfloor \nu/2 \rfloor} (-1)^k \frac{2^{\nu-2k}(n-k-1)!}{k!(\nu-2k)!} D^{\nu-2k} \delta(t), \qquad (26)$$
$$= T_n(D)\delta(t),$$

which are the distributional solutions of the forms (23). □

Example 5. *Letting $\nu = 1$, (22) becomes*

$$t^2 y''(t) + 3t y'(t) - t^2 y(t) = 0. \qquad (27)$$

From Theorem 2, (27) has a solution

$$y(t) = \delta'(t). \qquad (28)$$

Letting $\nu = 4$, (22) becomes

$$t^2 y''(t) + 2t y'(t) - (t^2 + 15)y(t) = 0. \qquad (29)$$

From Theorem 2, (29) has a solution

$$y(t) = 8\delta^{(4)}(t) - 8\delta''(t) + \delta(t). \qquad (30)$$

By applying (10), it is not difficult to verify that (28) and (30) satisfy (27) and (29), respectively.

4. Conclusions

In this paper, we seek the distributional solutions of the modified spherical Bessel differential Equation (11) and the linear differential equation of the form (22) by using the Laplace transforms of right-sided distributions and the power series solutions. The obtained solutions in the forms of the finite linear combinations of the Dirac delta function and its derivatives depend on the value of ν, to which their coefficients regard the coefficients of Legendre and Chebyshev polynomials (see [20] for more details). However, for solutions of (11) and (22) in the usual sense, not mentioned here, they can be seen in many standard and technical textbooks (see, for example, Ross [21]) but, even more, may appear in models related to equilibrium of membrane structures, steady states of evolutive equations or nonlinear science (see studies [22–25]).

Author Contributions: All authors contributed equally to this article. They read and approved the final manuscript. All authors have read and agreed to the published version of the manuscript.

Funding: This research received no external funding.

Acknowledgments: This work was supported by the Research Fund for Supporting Lecturer to Admit High Potential Student to Study and Research on His Expert Program Year 2018 from the Graduate School, Khon Kaen University, Thailand (Grant no. 611T105).

Conflicts of Interest: The authors declare no conflict of interest.

References

1. Wiener, J. Generalized-function solutions of differential and functional differential equations. *J. Math. Anal. Appl.* **1982**, *88*, 170–182. [CrossRef]
2. Kananthai, A. Distribution solutions of the third order Euler equation. *Southeast Asian Bull. Math.* **1999**, *23*, 627–631.
3. Liangprom, A.; Nonlaopon, K. On the generalized solutions of a certain fourth order Euler equations. *J. Nonlinear Sci. Appl.* **2017**, *10*, 4077–4084. [CrossRef]
4. Sacorn, N.; Nonlaopon, K.; Kim, H. A note on the generalized solutions of the third-order Cauchy-Euler equations. *Commun. Math. Appl.* **2018**, *9*, 661–669.
5. Jodnok, P.; Nonlaopon, K. On the generalized solutions of the fifth order Euler equations. *Far East J. Math. Sci.* **2018**, *106*, 59–74. [CrossRef]
6. Mirumbe, G.I.; Mango, J.M. On generalized solutions of locally Fuchsian ordinary differential equations. *J. Math. Sci. Adv. Appl.* **2018**, *51*, 99–117. [CrossRef]
7. Jhanthanam, S.; Nonlaopon, K.; Orankitjaroen, S. Generalized solutions of the third-order Cauchy-Euler equation in the space of right-sided distributions via Laplace transform. *Mathematics* **2019**, *7*, 376. [CrossRef]
8. Sangsuwan, A.; Nonlaopon, K.; Orankitjaroen, S.; Mirumbe, I. The generalized solutions of the nth order Cauchy-Euler equation. *Mathematics* **2019**, *7*, 932. [CrossRef]
9. Wiener, J. Generalized-function solutions of linear systems. *J. Differ. Equ.* **1980**, *38*, 301–315. [CrossRef]
10. Wiener, J.; Cooke, K.L.; Shah, S.M. Coexistence of analytic and distributional solutions for linear differential equations, II. *J. Math. Anal. Appl.* **1991**, *159*, 271–289. [CrossRef]
11. Hernandez, L.G.; Estrada, R. Solution of ordinary by series of delta functions. *J. Math. Anal. Appl.* **1995**, *191*, 40–55. [CrossRef]
12. Kanwal, R.P. Delta series solutions of differential and integral equations. *Integral Transform. Spec. Funct.* **1998**, *6*, 49–62. [CrossRef]
13. Nonlaopon, K.; Nuntigrangjana, T.; Putjuso, S. Distributional solutions of nth-order differential equations of the Bessel equation. *RMUTSB Acad. J.* **2017**, *5*, 1–10.
14. Kanwal, R.P. *Generalized Functions: Theory and Technique*, 3rd ed.; Springer: New York, NY, USA, 2004.
15. Cooke, K.L.; Wiener, J. Distributional and analytic solutions of functional differential equations. *J. Math. Anal. Appl.* **1984**, *98*, 111–129. [CrossRef]
16. Littlejohn, L.L.; Kanwal, R.P. Distributional solutions of the hypergeometric differential equation. *J. Math. Anal. Appl.* **1987**, *122*, 325–345. [CrossRef]
17. Wiener, J.; Cooke, K.L. Coexistence of analytic and distributional solutions for linear differential equations, I. *J. Math. Anal. Appl.* **1990**, *148*, 390–421. [CrossRef]
18. Boyvel, L.P.; Jones, A.R. *Electromagnetic Scattering and Its Applications*; Applied Science Publishers: London, UK, 1981.
19. Zemanian, A.H. *Distribution Theory and Transform Analysis*; McGraw-Hill: New York, NY, USA, 1965.
20. Wiener, J. *Generalized Solutions of Functional Differential Equations*; World Scientific: Singapore, 1993.
21. Ross, S.L. *Differential Equations*, 3rd ed.; John Wiley & Sons Inc.: Singapore, 1984.
22. Li, T.; Viglialoro, G. Analysis and explicit solvability of degenerate tensorial problems. *Bound. Value Probl.* **2018**, *1*, 1–13. [CrossRef]
23. Viglialoro, G.; Woolley, T.E. Boundedness in a parabolic-elliptic chemotaxis system with nonlinear diffusion and sensitivity and logistic source. *Math. Methods Appl. Sci.* **2018**, *41*, 1809–1824. [CrossRef]

24. Az-Zóbi, E.A.; Al-Khaled, K.; Darweesh, A. Numeric-analytic solutions for nonlinear oscillators via the modified multi-stage decomposition method. *Mathematics* **2019**, *7*, 550. [CrossRef]
25. González-Gaxiola, O.; Santiago, J.A.; Ruiz de Chávez, J. Solution for the nonlinear relativistic harmonic oscillator via Laplace-Adomian decomposition method. *Int. J. Appl. Comput. Math.* **2017**, *3*, 2627–2638. [CrossRef]

© 2020 by the authors. Licensee MDPI, Basel, Switzerland. This article is an open access article distributed under the terms and conditions of the Creative Commons Attribution (CC BY) license (http://creativecommons.org/licenses/by/4.0/).

Article

On the Solvability of Nonlinear Third-Order Two-Point Boundary Value Problems

Ravi P. Agarwal [1,*], Petio S. Kelevedjiev [2] and Todor Z. Todorov [2]

[1] Department of Mathematics, Texas A and M University-Kingsville, Kingsville, TX 78363-8202, USA
[2] Department of Mathematics, Technical University of Sliven, 8800 Sliven, Bulgaria; pskeleved@abv.bg (P.S.K.); tjtodorov@abv.bg (T.Z.T.)
* Correspondence: agarwal@tamuk.edu

Received: 21 April 2020; Accepted: 23 May 2020; Published: 31 May 2020

Abstract: Under barrier strips type assumptions we study the existence of $C^3[0,1]$—solutions to various two-point boundary value problems for the equation $x''' = f(t, x, x', x'')$. We give also some results guaranteeing positive or non-negative, monotone, convex or concave solutions.

Keywords: third-order differential equation; boundary value problem; existence; sign conditions

MSC: 34B15; 34B18

---

1. Introduction

In this paper, we study the solvability of boundary value problems (BVPs) for the differential equation

$$x''' = f(t, x, x', x''), t \in (0,1), \tag{1}$$

with some of the boundary conditions

$$x(0) = A, x'(1) = B, x''(1) = C, \tag{2}$$
$$x(0) = A, x'(0) = B, x''(1) = C, \tag{3}$$
$$x(0) = A, x(1) = B, x''(1) = C, \tag{4}$$
$$x(0) = A, x'(0) = B, x'(1) = C, \tag{5}$$
$$x(1) = A, x'(0) = B, x'(1) = C, \tag{6}$$

where $f : [0,1] \times D_x \times D_p \times D_q \to \mathbb{R}$, $D_x, D_p, D_q \subseteq \mathbb{R}$, and $A, B, C \in \mathbb{R}$.

The solvability of BVPs for third-order differential equations has been investigated by many authors. Here, we will cite papers devoted to two-point BVPs which are mostly with some of the above boundary conditions; in each of these works $A, B, C = 0$. Such problems for equations of the form

$$x''' = f(t, x), t \in (0, 1),$$

have been studied by H. Li et al. [1], S. Li [2] (the problem may be singular at $t=0$ and/or $t=1$), Z. Liu et al. [3,4], X. Lin and Z. Zhao [5], S. Smirnov [6], Q. Yao and Y. Feng [7]. Moreover, the boundary conditions in References [2,3] are (3), in Reference [4] they are (4), in References [1,5,7] they are (5), and in Reference [6] are

$$x(0) = x(1) = 0, x'(0) = C.$$

Y. Feng [8] and Y. Feng and S. Liu [9] have considered the equation

$$x''' = f(t, x, x'), t \in (0, 1),$$

with (6) and (5), respectively. Y. Feng [10] and R. Ma and Y. Lu [11] have studied the equations

$$f(t,x,x',x''') = 0 \text{ and } x''' + Mx'' + f(t,x) = 0, t \in (0,1).$$

with (5). BVPs for the equation

$$x''' = f(t,x,x',x''), t \in (0,1),$$

have been investigated by A. Granas et al. [12], B. Hopkins and N. Kosmatov [13], Y. Li and Y. Li [14]; the boundary conditions in [12] are (5), these in Reference [13] are (2) and (3), and in Reference [14]—(2).

Results guaranteeing positive or non-negative solutions can be found in References [2–4,7–11,13,14], and results that guarantee negative or non-positive ones in References [7,9,10]. The existence of monotone solutions has been studied in References [3,7,9].

As a rule, the main nonlinearity is defined and continuous on a set such that each dependent variable changes in a left- and/or a right-unbounded set; in Reference [13] it is a Carathéodory function on an unbounded set. Besides, the main nonlinearity is monotone with respect to some of the variables in References [1,5], does not change its sign in References [2–4,14] and satisfies Nagumo type growth conditions in Reference [14]. Maximum principles have been used in References [8,10], Green's functions in References [1,2,4,5], and upper and lower solutions in References [1,7–11].

Here, we use a different tool—barrier strips which allow the right side of the equation to be defined and continuous on a bounded subset of its domain and to change its sign.

To prove our existence results we apply a basic existence theorem whose formulation requires the introduction of the BVP

$$x''' + a(t)x'' + b(t)x' + c(t)x = f(t,x,x',x''), t \in (0,1), \tag{7}$$

$$V_i(x) = r_i, i = 1,2,3 (i = \overline{1,3} \text{ for short}), \tag{8}$$

where $a,b,c \in C([0,1],\mathbb{R}), f : [0,1] \times D_x \times D_p \times D_q \to \mathbb{R}$,

$$V_i(x) = \sum_{j=0}^{2}[a_{ij}x^{(j)}(0) + b_{ij}x^{(j)}(1)], i = \overline{1,3},$$

with constants a_{ij} and b_{ij} such that $\sum_{j=0}^{2}(a_{ij}^2 + b_{ij}^2) > 0, i = \overline{1,3}$, and $r_i \in \mathbb{R}, i = \overline{1,3}$. Next, consider the family of BVPs for

$$x''' + a(t)x'' + b(t)x' + c(t)x = g(t,x,x',x'',\lambda), t \in (0,1), \lambda \in [0,1] \tag{7}_\lambda$$

with boundary conditions (8), where g is a scalar function defined $[0,1] \times D_x \times D_p \times D_q \times [0,1]$, and a,b,c are as above. Finally, BC denotes the set of functions satisfying boundary conditions (8), and BC_0 denotes the set of functions satisfying the homogeneous boundary conditions $V_i(x) = 0, i = \overline{1,3}$. Besides, let $C^3_{BC}[0,1] = C^3[0,1] \cap BC$ and $C^3_{BC_0}[0,1] = C^3[0,1] \cap BC_0$.

The proofs of our existence results are based on the following theorem. It is a variant of Reference [12] (Chapter I, Theorem 5.1 and Chapter V, Theorem 1.2). Its proof can be found in Reference [15]; see also the similar result in Reference [16] (Theorem 4).

Lemma 1. *Suppose:*
(i) *Problem $(7)_0$, (8) has a unique solution $x_0 \in C^3[0,1]$.*
(ii) *Problems (7), (8) and $(7)_1$, (8) are equivalent.*
(iii) *The map $\mathbf{L}_h : C^3_{BC_0}[0,1] \to C[0,1]$ is one-to-one: here,*

$$\mathbf{L}_h x = x''' + a(t)x'' + b(t)x' + c(t)x.$$

(iv) Each solution $x \in C^3[0,1]$ to family $(7)_\lambda$, (8) satisfies the bounds

$$m_i \leq x^{(i)} \leq M_i \text{ for } t \in [0,1], i = \overline{0,3},$$

where the constants $-\infty < m_i, M_i < \infty, i = \overline{0,3}$, are independent of λ and x.

(v) There is a sufficiently small $\sigma > 0$ such that

$$[m_0 - \sigma, M_0 + \sigma] \subseteq D_x, [m_1 - \sigma, M_1 + \sigma] \subseteq D_p, [m_2 - \sigma, M_2 + \sigma] \subseteq D_q,$$

and $g(t, x, p, q, \lambda)$ is continuous for $(t, x, p, q, \lambda) \in [0, 1] \times J \times [0, 1]$ where $J = [m_0 - \sigma, M_0 + \sigma] \times [m_1 - \sigma, M_1 + \sigma] \times [m_2 - \sigma, M_2 + \sigma]$; $m_i, M_i, i = \overline{0,3}$, are as in (iv).

Then boundary value problem (7), (8) has at least one solution in $C^3[0,1]$.

For us, the equation from $(7)_\lambda$ has the form

$$x''' = \lambda f(t, x, x', x''). \qquad (1)_\lambda$$

Preparing the application of Lemma 1, we impose conditions which ensure the a priori bounds from (iv) for the eventual $C^3[0,1]$ - solutions of the families of BVPs for $(7)_\lambda, \lambda \in [0,1]$, with one of the boundary conditions $(k), k = \overline{2,6}$.

So, we will say that for some of the BVPs (1), $(k), k = \overline{2,6}$, the conditions $(\mathbf{H_1})$ and $(\mathbf{H_2})$ hold for a $K \in \mathbb{R}$ (it will be specified later for each problem) if:

$(\mathbf{H_1})$ There are constants $F'_i, L'_i, i = 1, 2$, such that

$$F'_2 < F'_1 \leq K \leq L'_1 < L'_2, [F'_2, L'_2] \subseteq D_q,$$

$$f(t, x, p, q) \geq 0 \text{ for } (t, x, p, q) \in [0,1] \times D_x \times D_p \times [L'_1, L'_2], \qquad (9)$$
$$f(t, x, p, q) \leq 0 \text{ for } (t, x, p, q) \in [0,1] \times D_x \times D_p \times [F'_2, F'_1]. \qquad (10)$$

$(\mathbf{H_2})$ There are constants $F_i, L_i, i = 1, 2$, such that

$$F_2 < F_1 \leq K \leq L_1 < L_2, [F_2, L_2] \subseteq D_q,$$

$$f(t, x, p, q) \leq 0 \text{ for } (t, x, p, q) \in [0,1] \times D_x \times D_p \times [L_1, L_2],$$
$$f(t, x, p, q) \geq 0 \text{ for } (t, x, p, q) \in [0,1] \times D_x \times D_p \times [F_2, F_1].$$

Besides, we will say that for some of the BVPs (1), $(k), k = \overline{2,6}$, the condition $(\mathbf{H_3})$ holds for constants $m_i \leq M_i, i = \overline{0,2}$, (they also will be specified later for each problem) if:

$(\mathbf{H_3})$ $[m_0 - \sigma, M_0 + \sigma] \subseteq D_x, [m_1 - \sigma, M_1 + \sigma] \subseteq D_p, [m_2 - \sigma, M_2 + \sigma] \subseteq D_q$ and $f(t, x, p, q)$ is continuous on the set $[0,1] \times J$, where J is as in (v) of Lemma 1, and $\sigma > 0$ is sufficiently small.

In fact, the present paper supplements P. Kelevedjiev and T. Todorov [15] where only conditions $(\mathbf{H_2})$ and $(\mathbf{H_3})$ have been used for studying the solvability of various BVPs for (1) with other boundary conditions. Here, $(\mathbf{H_1})$ is also needed. Now, only $(\mathbf{H_1})$ guarantees the a priori bounds for $x''(t), x'(t)$ and $x(t)$, in this order, for each eventual solution $x \in C^3[0,1]$ to the families $(1)_\lambda, (k), k = \overline{2,4}$, and $(\mathbf{H_1})$ and $(\mathbf{H_2})$ together guarantee these bounds for the families $(1)_\lambda, (k), k = 5, 6$. As in Reference [15], $(\mathbf{H_3})$ gives the bounds for $x'''(t)$.

The auxiliary results which guarantee a priori bounds are given in Section 2, and the existence theorems are in Section 3. The ability to use $(\mathbf{H_1})$ and $(\mathbf{H_2})$ for studying the existence of solutions with important properties is shown in Appendix A. Examples are given in Section 4.

2. Auxiliary Results

This part ensures a priori bounds for the eventual $C^3[0, 1]$-solutions of each family $(1)_\lambda$, (k), $k = \overline{2, 6}$, that is, it ensures the constants m_i, M_i, $i = \overline{0, 2}$, from *(iv)* of Lemma 1 and **(H$_3$)**.

Lemma 2. *Let $x \in C^3[a, b]$ be a solution to $(1)_\lambda$. Suppose* **(H$_1$)** *holds with $[0, 1]$ replaced by $[a, b]$ and $K = x''(b)$. Then*
$$F_1' \leq x''(t) \leq L_1' \text{ on } [a, b].$$

Proof. By contradiction, assume that $x''(t) > L_1'$ for some $t \in [a, b)$. This means that the set
$$S_+ = \{t \in [a, b] : L_1' < x''(t) \leq L_2'\}$$
is not empty because $x''(t)$ is continuous on $[a, b]$ and $x''(b) \leq L_1'$. Besides, there is a $\gamma \in S_+$ such that
$$x'''(\gamma) < 0.$$

As $x(t)$ is a $C^3[a, b]$—solution to $(1)_\lambda$,
$$x'''(\gamma) = \lambda f(\gamma, x(\gamma), x'(\gamma), x''(\gamma)).$$

But, $(\gamma, x(\gamma), x'(\gamma), x''(\gamma)) \in S_+ \times D_x \times D_p \times (L_1', L_2']$ and (9) imply
$$x'''(\gamma) \geq 0,$$
a contradiction. Consequently,
$$x''(t) \leq L_1' \text{ for } t \in [a, b].$$

Along similar lines, assuming on the contrary that the set
$$S_- = \{t \in [a, b] : F_2' \leq x''(t) < F_1'\}$$
is not empty and using (10), we achieve a contradiction which implies that
$$F_1' \leq x''(t) \text{ for } t \in [a, b].$$

□

The proof of the next assertion is virtually the same as that of Lemma 2 and is omitted; it can be found in [15].

Lemma 3. *Let $x \in C^3[a, b]$ be a solution to $(1)_\lambda$. Suppose* **(H$_2$)** *holds with $[0, 1]$ replaced by $[a, b]$ and $K = x''(a)$. Then*
$$F_1 \leq x''(t) \leq L_1 \text{ on } [a, b].$$

Let us recall, conditions of type **(H$_1$)** and **(H$_2$)** are called barrier strips, see P. Kelevedjiev [17]. As can we see from Lemmas 2 and 3 they control the behavior of $x''(t)$ on $[a, b]$, depending on the sign of $f(t, x, x', x'')$ the curve of $x''(t)$ on $[a, b]$ crosses the strips $[a, b] \times [L_1', L_2']$, $[a, b] \times [L_1, L_2]$, $[a, b] \times [F_2', F_1']$ and $[a, b] \times [F_2, F_1]$ not more than once. This property ensures the a priori bounds for $x''(t)$.

Lemma 4. *Let* **(H$_1$)** *hold for $K = C$. Then every solution $x \in C^3[0, 1]$ to $(1)_\lambda$, (2) or $(1)_\lambda$, (3) satisfies the bounds*
$$|x(t)| \leq |A| + |B| + \max\{|F_1'|, |L_1'|\}, t \in [0, 1],$$
$$|x'(t)| \leq |B| + \max\{|F_1'|, |L_1'|\}, t \in [0, 1],$$

$$F_1' \leq x''(t) \leq L_1', t \in [0,1]. \tag{11}$$

Proof. Let first $x(t)$ be a solution to $(1)_\lambda$, (2). Using Lemma 2 we conclude that (11) is true. Then, according to the mean value theorem, for each $t \in [0,1)$ there is a $\xi \in (t,1)$ such that

$$x'(1) - x'(t) = x''(\xi)(1-t),$$

which together with (11) gives the bound for $|x'(t)|$. Again from the mean value theorem for each $t \in (0,1]$ there is an $\eta \in (0,t)$ with the property

$$x(t) - x(0) = x'(\eta)t,$$

which yields the bound for $|x(t)|$. The assertion follows similarly for $(1)_\lambda$, (3). □

Lemma 5. *Let* **(H$_1$)** *hold for* $K = C$. *Then every solution* $x \in C^3[0,1]$ *to* $(1)_\lambda$, (4) *satisfies the bounds*

$$|x(t)| \leq |A| + |B - A| + \max\{|F_1'|, |L_1'|\}, \ t \in [0,1],$$

$$|x'(t)| \leq |B - A| + \max\{|F_1'|, |L_1'|\}, \ t \in [0,1],$$

$$F_1' \leq x''(t) \leq L_1', t \in [0,1].$$

Proof. By Lemma 2, $F_1' \leq x''(t) \leq L_1'$ on $[0,1]$. Clearly, there is a $\mu \in (0,1)$ for which $x'(\mu) = B - A$. Further, for each $t \in [0,\mu)$ there is a $\xi \in (t,\mu)$ such that

$$x'(\mu) - x'(t) = x''(\xi)(\mu - t),$$

from where, using the obtained bounds for $x''(t)$, we get

$$|x'(t)| \leq |B - A| + \max\{|F_1'|, |L_1'|\}, \ t \in [0,\mu].$$

We can proceed analogously to see that the same bound is valid for $t \in [\mu, 1]$. Finally, for each $t \in (0,1]$ there is an $\eta \in (0,t)$ such that

$$x(t) - x(0) = x'(\eta)t,$$

which together with the obtained bound for $|x'(t)|$ yields the bound for $|x(t)|$. □

Lemma 6. *Let* **(H$_1$)** *and* **(H$_2$)** *hold for* $K = C - B$. *Then every solution* $x \in C^3[0,1]$ *to* $(1)_\lambda$, (5) *or* $(1)_\lambda$, (6) *satisfies the bounds*

$$|x(t)| \leq |A| + |B| + \max\{|F_1|, |L_1|, |F_1'|, |L_1'|\}, t \in [0,1],$$

$$|x'(t)| \leq |B| + \max\{|F_1|, |L_1|, |F_1'|, |L_1'|\}, t \in [0,1],$$

$$\min\{F_1, F_1'\} \leq x''(t) \leq \max\{L_1, L_1'\}, t \in [0,1].$$

Proof. Let $x(t)$ be a solution to $(1)_\lambda$, (5); the proof is similar for $(1)_\lambda$, (6). We know there is a $\nu \in (0,1)$ for which $x''(\nu) = C - B$. Then, applying Lemmas 2 and 3 on the intervals $[0,\nu]$ and $[\nu,1]$, respectively, we get

$$F_1' \leq x''(t) \leq L_1' \text{ on } [0,\nu] \text{ and } F_1 \leq x''(t) \leq L_1 \text{ on } [\nu,1]$$

and so the bounds for $x''(t)$ follow. Further, as in the proof of Lemma 4 we establish consecutively the bounds for $|x'(t)|$ and $|x(t)|$. □

3. Existence Results

Theorem 1. *Let* **(H$_1$)** *hold for* $K = C$ *and* **(H$_3$)** *hold for*

$$M_0 = |A| + |B| + \max\{|F'_1|, |'L'_1|\}, m_0 = -M_0,$$

$$M_1 = |B| + \max\{|F'_1|, |L'_1|\}, m_1 = -M_1, m_2 = F'_1, M_2 = L'_1.$$

Then each of BVPs (1), (2) *and* (1), (3) *has at least one solution in* $C^3[0,1]$.

Proof. We will establish that the assertion is true for problem (1), (2) after checking that the hypotheses of Lemma 1 are fulfilled; it follows similarly and for (1), (3). We easily check that *(i)* holds for (1)$_0$, (2). Clearly, BVP (1), (2) is equivalent to BVP (1)$_1$, (2) and so *(ii)* is satisfied. Since now $\mathbf{L}_h = x'''$, *(iii)* also holds. Next, according to Lemma 4, for each solution $x \in C^3[0,1]$ to (1)$_\lambda$, (2) we have

$$m_i \leq x^{(i)}(t) \leq M_i, t \in [0,1], i = 0, 1, 2.$$

Now use that f is continuous on $[0,1] \times J$ to conclude that there are constants m_3 and M_3 such that

$$m_3 \leq \lambda f(t, x, p, q) \leq M_3 \text{ for } \lambda \in [0,1] \text{ and } (t, x, p, q) \in [0,1] \times J,$$

which together with $(x(t), x'(t), x''(t)) \in J$ for $t \in [0,1]$ and Equation (1)$_\lambda$ implies

$$m_3 \leq x'''(t) \leq M_3, t \in [0,1].$$

These observations imply that *(iv)* holds, too. Finally, the continuity of f on the set J gives *(v)* and so the assertion is true by Lemma 1. □

Theorem 2. *Let* **(H$_1$)** *hold for* $K = C$ *and* **(H$_3$)** *hold for*

$$M_0 = |A| + |B - A| + \max\{|F'_1|, |L'_1|\}, m_0 = -M_0,$$

$$M_1 = |B - A| + \max\{|F'_1|, |L'_1|\}, m_1 = -M_1, m_2 = F'_1, M_2 = L'_1.$$

Then BVP (1), (4) *has at least one solution in* $C^3[0,1]$.

Proof. It follows the lines of the proof of Theorem 1. Now the bounds

$$m_i \leq x^{(i)}(t) \leq M_i, t \in [0,1], i = 0, 1, 2,$$

for each solution $x \in C^3[0,1]$ to a (1)$_\lambda$, (4) follow from Lemma 5. □

Theorem 3. *Let* **(H$_1$)** *and* **(H$_2$)** *hold for* $K = C - B$ *and* **(H$_3$)** *hold for*

$$M_0 = |A| + |B| + \max\{|F_1|, |L_1|, |F'_1|, |L'_1|\}, m_0 = -M_0,$$

$$M_1 = |B| + \max\{|F_1|, |L_1|, |F'_1|, |L'_1|\}, m_1 = -M_1,$$

$$m_2 = \min\{F_1, F'_1\}, M_2 = \max\{L_1, L'_1\}.$$

Then each of BVPs (1), (5) *and* (1), (6) *has at least one solution in* $C^3[0,1]$.

Proof. Arguments similar to those in the proof of Theorem 1 yield the assertion. Now the bounds

$$m_i \leq x^{(i)}(t) \leq M_i, t \in [0,1], i = 0, 1, 2,$$

for each solution $x \in C^3[0,1]$ to $(1)_\lambda$, (5) and $(1)_\lambda$, (6) follow from Lemma 6. □

4. Examples

Through several examples we will illustrate the application of the obtained results.

Example 1. *Consider the BVPs for the equation*

$$x'''(t) = \exp(x'' - 3) + 5x''(x'^2 + 1) - t\sin x, t \in (0,1),$$

with boundary conditions (2) *or* (3).

For $F_2' = -|C| - 2$, $F_1' = -|C| - 1$, $L_1' = \max\{|C|, 3\} + 1$, $L_2' = \max\{|C|, 3\} + 2$ and $\sigma = 0.1$, for example, each of these problems has a solution in $C^3[0,1]$ by Theorem 1.

Example 2. *Consider the BVP*

$$x'''(t) = \varphi(t, x, x')\Big(\lg((x'' + 50)(60 - x'')) - 3\Big), t \in (0,1),$$

$$x(0) = 5, x'(0) = 10, x'(1) = 40,$$

where $\varphi : [0,1] \times \mathbb{R}^2 \to \mathbb{R}$ *is continuous and does not change its sign.*

If $\varphi(t,x,p) \geq 0$ on $[0,1] \times \mathbb{R}^2$, the assumptions of Theorem 3 are satisfied for $F_2 = -36$, $F_1 = -35$, $F_2' = -46$, $F_1' = -45$, $L_1' = 40$, $L_2' = 41$, $L_1 = 55$, $L_2 = 56$ and $\sigma = 0.01$, for example, and if $\varphi(t,x,p) \leq 0$ on $[0,1] \times \mathbb{R}^2$, they are satisfied for $F_2' = -36$, $F_1' = -35$, $F_2 = -46$, $F_1 = -45$, $L_1 = 40$, $L_2 = 41$, $L_1' = 55$, $L_2' = 56$ and $\sigma = 0.01$, for example; it is clear, $K = 30$. Thus, the considered problem has at least one solution in $C^3[0,1]$. Let us note, here $D_q = (-50, 60)$.

Example 3. *Consider the BVP*

$$x'''(t) = \frac{t(x''+8)(x''+3)\sqrt{625 - x'^2}}{\sqrt{900 - x^2}\sqrt{100 - x''^2}}, t \in (0,1),$$

$$x(0) = 9, x(1) = 1, x''(1) = -4.$$

For $F_2' = -6$, $F_1' = -5$, $L_1' = -3$, $L_2' = -2$ and $\sigma = 0.1$, for example, this problem has a positive, decreasing, concave solution in $C^3[0,1]$ by Theorem A1; notice, here D_x, D_p and D_q are bounded.

Author Contributions: All authors contributed equally. All authors have read and agreed to the published version of the manuscript.

Funding: This research received no external funding.

Conflicts of Interest: The authors declare no conflict of interest.

Appendix A

In this part we show how the barrier strips can be used for studying the existence of positive or non-negative, monotone, convex or concave $C^3[0,1]$ - solutions. Here, we demonstrate this on problem (1), (4) but it can be done for the rest of the BVPs considered in this paper. Similar results for various other two-point boundary conditions can be found in R. Agarwal and P. Kelevedjiev [16] and P. Kelevedjiev and T. Todorov [15].

Lemma A1. Let $A, B \geq 0, C \leq 0$. Suppose **(H$_1$)** holds for $K = C$ with $L'_1 \leq 0$. Then each solution $x \in C^3[0, 1]$ to $(1)_\lambda$, (4) satisfies the bounds

$$\min\{A, B\} \leq x(t) \leq A + |B - A| + |F'_1|, \ t \in [0, 1],$$

$$B - A + F'_1 \leq x'(t) \leq B - A - F'_1, \ t \in [0, 1].$$

Proof. From Lemma 2 we know that $F'_1 \leq x''(t) \leq L'_1$ for $t \in [0, 1]$. Besides, for some $\mu \in (0, 1)$ we have $x'(\mu) = B - A$. Then,

$$\int_t^\mu F'_1 ds \leq \int_t^\mu x''(s) ds \leq \int_t^\mu L'_1 ds, t \in [0, \mu),$$

gives

$$B - A \leq x'(t) \leq B - A - F'_1, t \in [0, \mu],$$

and

$$\int_\mu^t F'_1 ds \leq \int_\mu^t x''(s) ds \leq \int_\mu^t L'_1 ds, t \in (\mu, 1],$$

implies

$$B - A + F'_1 \leq x'(t) \leq B - A, t \in [\mu, 1].$$

As a result,

$$B - A + F'_1 \leq x'(t) \leq B - A - F'_1, t \in [0, 1].$$

Using Lemma 5, conclude

$$|x(t)| \leq A + |B - A| + |F'_1| \text{ for } t \in [0, 1].$$

From $x''(t) \leq L'_1 \leq 0$ for $t \in [0, 1]$ it follows that $x(t)$ is concave on $[0, 1]$ and so, in view of $A, B \geq 0$, $x(t) \geq \min\{A, B\}$ on $[0, 1]$, which completes the proof. $\square$

Theorem A1. Let $A \geq B \geq 0$ and $C \leq 0$ ($A \geq B > 0$ and $C < 0$). Suppose **(H$_1$)** holds for $K = C$ with $B - A \leq F'_1$ ($B - A < F'_1$) and $L'_1 \leq 0$, and **(H$_3$)** holds for

$$m_0 = B, M_0 = 2A - B + |F'_1|,$$

$$m_1 = B - A + F'_1, M_1 = B - A - F'_1, m_2 = F'_1, M_2 = L'_1.$$

Then BVP (1), (4) has at least one non-negative, non-increasing (positive, decreasing), concave solution in $C^3[0, 1]$.

Proof. By Lemma 5, for every solution $x \in C^3[0, 1]$ to $(1)_\lambda$, (4) we have $F'_1 \leq x''(t) \leq L'_1$ on $[0, 1]$, and Lemma A1 yields

$$B - A + F'_1 \leq x'(t) \leq B - A - F'_1, \ t \in [0, 1]$$

$$\min\{A, B\} \leq x(t) \leq A + |B - A| + |F'_1|, \ t \in [0, 1].$$

Because of $A \geq B$, the last inequality gets the form

$$B \leq x(t) \leq 2A - B + |F'_1|, \ t \in [0, 1].$$

So, $x(t)$ satusfies the bounds

$$m_0 \leq x^{(i)}(t) \leq M_0, t \in [0, 1], i = 0, 1, 2.$$

Essentially the same reasoning as in the proof of Theorem 1 establishes that (1), (4) has a solution in $C^3[0,1]$. Since $m_0 = B \geq 0 (m_0 > 0)$, $M_1 = B - A - F_1' \leq 0 (M_1 < 0)$ and $M_2 = L_1' \leq 0$, this solution has the desired properties. □

Reference

1. Li, H.; Feng, Y.; Bu, C. Non-conjugate boundary value problem of a third order differential equation. *Electron. J. Qual. Theory Differ. Equ.* **2015**, *2015*, 1–19. [CrossRef]
2. Li, S. Positive solutions of nonlinear singular third-order two-point boundary value problem. *J. Math. Anal. Appl.* **2006**, *323*, 413–425. [CrossRef]
3. Liu, Z.; Debnath, L.; Kang, S.M. Existence of monotone positive solutions to a third order two-point generalized right focal boundary value problem. *Comput. Math. Appl.* **2008**, *55*, 356–367. [CrossRef]
4. Liu, Z.; Ume, J.S.; Kang, S.M. Positive solutions of a singular nonlinear third order two-point boundary value problem. *J. Math. Anal. Appl.* **2007**, *326*, 589–601. [CrossRef]
5. Lin, X.; Zhao, Z. Sign-changing solution for a third-order boundary-value problem in ordered Banach space with lattice structure. *Bound. Value Probl.* **2014**, *2014*, 132. [CrossRef]
6. Smirnov, S. On the third order boundary value problems with asymmetric nonlinearity. *Nonlinear Anal. Model. Control* **2011**, *16*, 231–241. [CrossRef]
7. Yao, Q.; Feng, Y. The existence of solutions for a third-order two-point boundary value problem. *Appl. Math. Lett.* **2002**, *15*, 227–232. [CrossRef]
8. Feng, Y. Solution and positive solution of a semilinear third-order equation. *J. Appl. Math. Comput.* **2009**, *29*, 153–161. [CrossRef]
9. Feng, Y.; Liu, S. Solvability of a third-order two-point boundary value problem. *Appl. Math. Lett.* **2005**, *18*, 1034–1040. [CrossRef]
10. Feng, Y. Existence and uniqueness results for a third-order implicit differential equation. *Comput. Math. Appl.* **2008**, *56*, 2507–2514. [CrossRef]
11. Ma, R.; Lu, Y. Disconjugacy and extremal solutions of nonlinear third-order equations. *Commun. Pure Appl. Anal.* **2014**, *13*, 1223–1236. [CrossRef]
12. Granas, A.; Guenther, R.B.; Lee, J.W. *Nonlinear Boundary Value Problems for Ordinary Differential Equations*; Dissnes Math.: Warszawa, Poland, 1985.
13. Hopkins, B.; Kosmatov, N. Third-order boundary value problems with sign-changing solutions. *Nonlinear Anal.* **2007**, *67*, 126–137. [CrossRef]
14. Li, Y.; Li, Y. Positive solutions of a third-order boundary value problem with full nonlinearity. *Mediterr. J. Math.* **2017**, *14*, 128. [CrossRef]
15. Kelevedjiev, P.; Todorov, T.Z. Existence of solutions of nonlinear third-order two-point boundary value problems. *Electron. J. Qual. Theory Differ. Equ.* **2019**, *2019*, 1–15. [CrossRef]
16. Agarwal, R.P.; Kelevedjiev, P.S. On the solvability of fourth-order two-point boundary value problems. *Mathematics* **2020**, *8*, 603. [CrossRef]
17. Kelevedjiev, P. Existence of solutions for two-point boundary value problems *Nonlinear Anal.* **1994**, *22*, 217–224. [CrossRef]

© 2020 by the authors. Licensee MDPI, Basel, Switzerland. This article is an open access article distributed under the terms and conditions of the Creative Commons Attribution (CC BY) license (http://creativecommons.org/licenses/by/4.0/).

Article

On the Uniqueness Classes of Solutions of Boundary Value Problems for Third-Order Equations of the Pseudo-Elliptic Type

Abdukomil Risbekovich Khashimov [1],* and Dana Smetanová [2],*

[1] Tashkent Finance Institute, Tashkent 1000000, Uzbekistan
[2] Institute of Technology and Business in České Budějovice, 370 01 České Budějovice, Czech Republic
* Correspondence: abdukomil@yandex.ru (A.R.K.); smetanova@mail.vstecb.cz (D.S.)

Received: 9 June 2020; Accepted: 13 July 2020; Published: 16 July 2020

Abstract: The paper is devoted to solutions of the third order pseudo-elliptic type equations. An energy estimates for solutions of the equations considering transformation's character of the body form were established by using of an analog of the Saint-Venant principle. In consequence of this estimate, the uniqueness theorems were obtained for solutions of the first boundary value problem for third order equations in unlimited domains. The energy estimates are illustrated on two examples.

Keywords: equations of the pseudo-elliptic type of third order; energy estimate; analog of the Saint-Venant principle

PACS: 02.30.Jr

MSC: 35M20; 35Q99

1. Introduction

In the 19th century, A.J.C. Barré de Saint-Venant studied the planar theory of elasticity. His principle is expressed as a prior estimate for a solution of a biharmonic equation satisfying homogeneous boundary conditions of the first boundary value problem in the part of the domain boundary (c.f., [1,2]). Many recent recent results are inspired by Saint-Venant principle (c.f., [3–5] and many others).

The energetic estimates were received first in [6,7]. These estimates do not take into account character of transformation of the body form at moving off from those part of the bound where exterior forces are applied. In the paper [8], a proof of the Saint-Venant principle in the planar theory of elasticity was obtained by different way. The energetic estimate was gained in the connection considered character of transformation of the body form. The uniqueness theorem for the first boundary value problem of the planar theory of elasticity in unlimited domains and also Pharagmen–Lindelöf type theorems were obtained as a corollary of the energetic estimate. The proofs of the Pharagmen–Lindelöf type theorems were done for equations of the theory of elasticity in [9] and for elliptic equations of higher order in the papers [2,6,7,10–14]. The Saint-Venant principle for a cylindrical body was studied in [15].

Boundary value problems have applications in fluid dynamics, astrophysics, hydrodynamic, hydromagnetic stability, astronomy, beam and long wave theory, induction motors, engineering, and applied physics. Boundary value problems of higher order is studied in papers [16,17]. An overview of some results on the class of functions with subharmonic behaviour and their invariance properties under conformal and quasiconformal mappings is presented in [18].

An analog of the Saint-Venant principle, uniqueness theorems in unlimited domains, and Pharagmen–Lindelöf type theorems in the theory of elasticity were derived for the system of equations in the case of space with boundary conditions of the first boundary value problem (c.f., [19,20]). Similar results were obtained for the mixed problems in [21].

We shall note else work [12,22], which by means of principle Saint-Venant's is studied asymptotic characteristic of the solutions of the third order equations of the composite type and dynamic systems.

Boundary value problems have applications in fluid dynamics, astrophysics, hydrodynamic, hydromagnetic stability, astronomy, beam and long wave theory, induction motors, engineering, and applied physics.

2. Notations and Formulation of the Problem

Consider in the unlimited domain Q the equation

$$L_0 l u + L_1 u + M u = f(x,y,t) \tag{1}$$

where

$$l u = u_t + \alpha^k(x) u_{x_k} + \alpha_0(x) u, \quad L_1 u = b^{ij}(x) u_{x_i x_j} + b^i(x) u_{x_i},$$

$$L_0 u = u_t - a^{ij}(x) u_{x_i x_j} + a^i(x) u_{x_i} + a_0(x) u,$$

$$M u = c^{pq}(x) u_{y_p y_q} + c^p(x) u_{y_p} + c_0(x) u.$$

We suppose here and later on that the summation is carried out by repeating indexes, all coefficients in (1) and their derivatives are bounded and measurable in any finite subdomain of the domain Q. Furthermore, we suppose that boundary of Q is smooth or piecewise-smooth. We assume that the operators L_0, M are uniformly elliptic, i.e.,

$$a^{ij} = a^{ji}, \quad \lambda_0 |\xi|^2 \leq a^{ij} \xi_i \xi_j \leq \lambda_1 |\xi|^2, \quad \text{for all} \quad (x,y,t) \in Q \cup \partial Q, \quad \text{for all} \quad \xi \in \mathbb{R}^{n+m+1}$$

$$c^{pq} = c^{qp}, \quad \mu_0 |\xi|^2 \leq a^{ij} \xi_i \xi_j \leq \mu_1 |\xi|^2, \quad \text{for all} \quad (x,y,t) \in Q \cup \partial Q, \quad \text{for all} \quad \xi \in \mathbb{R}^{n+m+1}. \tag{2}$$

Let $G = D \times \Omega$ and $\nu(x) = (\nu_{x_1}, \ldots, \nu_{x_n}, \nu_{y_1}, \ldots, \nu_{y_m}, \nu_t)$ is a vector of the inner normal of Q in the point (x,y,t).

We break up the bound of Q. Denote

$$\sigma_0 = \{(x,y,t) \in \partial G \times (0,T) : \alpha^k \nu_k = 0\},$$

$$\sigma_1 = \{(x,y,t) \in \partial G \times (0,T) : \alpha^k \nu_k > 0\},$$

$$\sigma_2 = \{(x,y,t) \in \partial G \times (0,T) : \alpha^k \nu_k < 0\},$$

Consider in Q the boundary value problem

$$L_0 l u + L_1 u + M u = f(x,y,t),$$

$$u|_{\partial Q} = 0, \quad \alpha^k u_{x_k}|_{\sigma_2} = 0. \tag{3}$$

Define the operator d:

$$du = (b^{ij} + \alpha^k a^{ij}_{x_k} - \alpha_0 a^{ij} + a^{ij}_t) u_{x_i x_j} + (b^i + \alpha_0 a^i - \alpha^i a^k_{x_k} + \alpha^i a_0 - a^i_t) u_{x_i} + (a_{0_t} - \alpha_0 a_0) u \equiv$$

$$d^{ij} u_{x_i x_j} + d^i u_{x_i} + du.$$

Assume that the condition

$$d^{ij} = d^{ji}, \quad \gamma_0 |\xi|^2 \leq d^{ij} \xi_i \xi_j \leq \gamma_1 |\xi|^2, \quad \text{for all} \quad (x,y,t) \in Q \cup \partial Q, \quad \text{for all} \quad \xi \in \mathbb{R}^{n+m+1} \tag{4}$$

holds.

Let
$$Q_\tau = Q \cap \{(x,y,t) : 0 < y_1 < \tau\}, \quad \partial G_\tau = \partial G \cap \{y : 0 < y_1 < \tau\},$$
$$\sigma_{0,\tau} = \{(x.y.t) \in \partial G_\tau \times (0,T) : \alpha^k \nu_k = 0\},$$
$$\sigma_{1,\tau} = \{(x,y,t) \in \partial G_\tau \times (0,T) : \alpha^k \nu_k > 0\},$$
$$\sigma_{2,\tau} = \{(x,y,t) \in \partial G_\tau \times (0,T) : \alpha^k \nu_k < 0\}.$$

For some $h > 0$, define
$$\sigma_{2,h,\tau} = \{(x,y,t) \in \sigma_{2,\tau} : \rho((x,y,t), \partial \sigma_{2,\tau}) > h\}, \quad \sigma_{2,\tau}^h = \sigma_{2,\tau} \setminus \sigma_{2,h,\tau}.$$

Let $E(Q_\tau)$ be a set of functions $v \in C^2(\overline{Q}_\tau)$ such that $v = 0$ in $\partial G_\tau \times (0,T)$ and $\alpha^k v_{x_k} = 0$ on $\sigma_{0,\tau} \cup \sigma_{1,\tau} \cup \sigma_{2,\tau}^h$ for some $h > 0$.

We denote as $H(Q_\tau)$ the Hilbert space obtained by closing $E(Q_\tau)$ with respect to the norm

$$\|u\|_{H(Q_\tau)} = \left\{ \int_{Q_\tau} \left(d_1^{ij} u_{x_i} u_{x_j} + u_{y_p} u_{y_q} + u_t^2 + u^2 \right) dx\, dy\, dt - \int_{\sigma_{2,\tau}} \alpha^k \nu_k a^{ij} u_{x_i} u_{x_j} ds \right\}^{\frac{1}{2}},$$

where
$$d_1^{ij} = -\frac{1}{2} \alpha^j a_{x_j}^{ij} - \frac{1}{2} a_t^{ij} + \alpha^j a^i + d^{ij} - \frac{1}{2\lambda_0} a^{ij},$$

$d_1^{ij} = d_1^{ji}, \quad \beta_0 |\xi|^2 \le d_1^{ij} \xi_i \xi_j \le \beta_1 |\xi|^2$, for all $(x,y,t) \in Q \cup \partial Q$, for all $\xi \in \mathbb{R}^{n+n+1}$.

Now consider bilinear form
$$a(u,v) = \int_{Q_\tau} \left[\alpha^k a^{ij} u_{x_i} v_{x_j x_k} + a^{ij} u_{x_i} v_{x_j t} + \left(\alpha^k a_{x_j}^{ij} - \alpha^i a^k \right) u_{x_i} v_{x_j} + \right.$$
$$d^{ij} u_{x_i} v_{x_j} + \left(d^i - d_{x_j}^{ij} \right) uv_{x_i} + \left(a_{x_i}^{ij} + a^i + \alpha^i \right) u_{x_i} v_t + c^{pq} u_{y_p} v_{y_q} + \left(c^p - c_{y_q}^{pq} \right) uv_{y_p} +$$
$$\left. u_t v_t + (\alpha_0 + a_0) u v_t + \left(c_{y_p}^p - c_0 - c_{y_p y_q}^{pq} + d + d_{x_i}^i + d_{x_i x_j}^{ij} \right) uv\, dx\, dy\, dt. \right.$$

Definition 1. *If $u(x,y,t) \in H(Q_\tau)$ for any $\tau < +\infty$ and*
$$a(u,v) = \int_{Q_\tau} fv\, dx\, dy\, dt \tag{5}$$

for an arbitrary function $v \in E(Q_\tau)$, $v|_{S_\tau} = 0$ where $S_\tau = Q \cap \{(x,y,t) : y_1 = \tau\}$, then the function $u(x,y,t)$ is said to be a generalized solution of the problem (1),(3) in the domain Q.

3. Energy Inequalities

Theorem 1. *(Analog of the Saint-Venant principle)*
Let $-1 \le a_{x_i}^{ij} + a^i + a_0 \le 0$; $\theta \equiv d_0 - \frac{1}{2} d_{x_i x_j}^{ij} + \frac{1}{2} d_{x_i}^i - \frac{1}{2} c_{y_p y_q}^{pq} + \frac{1}{2} c_{y_p}^p - c_0 \le \theta_0 < 0$, for all $(x,y,t) \in Q \cup \partial Q$.

If $u(x,y,t)$ is generalized solution of the problem (1), (3) and $f(x,y,t) = 0$ at $y_1 \le \tau_2$, then for any τ_1 such that $0 \le \tau_1 \le \tau_2$, takes place

$$\int_{Q_{\tau_1}} E(u) dx\, dy\, dt \le \Phi^{-1}(\tau_1, \tau_2) \int_{Q_{\tau_2}} E(u) dx\, dy\, dt \tag{6}$$

where $E(u) = d^{ij} u_{x_i} u_{x_j} + c^{pq} u_{y_p} u_{y_q} + u_t^2 - \theta u^2$.

Here $\Phi(\tau,\tau_2)$ is a solution of the problem

$$\Phi' = -\mu(\tau)\Phi, \quad \tau_1 \leq \tau \leq \tau_2, \tag{7}$$
$$\Phi(\tau_2,\tau_2) = 1,$$

$\mu(\tau)$ is an arbitrary continuous function such that

$$0 < \mu(\tau) \leq \inf_N \left\{ \int_{S_\tau} E(v) dx\, dy'\, dt \left| \int_{S_\tau} P(v) dx\, dy'\, dt \right|^{-1} \right\}, \tag{8}$$

$$y' = (y_2, y_3, \ldots, y_m),$$
$$P(v) = -c^{p1} v v_{y_p} + \frac{1}{2}\left(c^1 - c_{y_q}^{1q}\right) v^2, \tag{9}$$

N is the set of continuously differentiable functions in the neighborhood of $\overline{S_\tau}$ which are equal to zero in $\overline{S_\tau} \cap (\partial G_\tau \times (0,T))$.

Proof. Assume in (5) $v = u_m(\Psi(y_1) - 1)$ where $\Psi(y_1) = \Phi(\tau_1, \tau_2)$ if $0 \leq y_1 \leq \tau_1$, $\Psi(y_1) = \Phi(y_1, \tau_2)$ if $\tau_1 \leq y_1 \leq \tau_2$, and $\Psi(y_1) = 1$ if $\tau_2 \leq y_1$.

$$u_m \in E(Q_\tau), \quad \|u_m - u\|_{H(Q_\tau)} \to 0, \quad u \in H(Q).$$

Then

$$a(u - u_m + u_m, u_m(\Psi - 1)) = 0 \text{ in } Q_{\tau_2}.$$

Therefore

$$a(u_m, u_m(\Psi - 1)) = \delta_m \text{ in } Q_{\tau_2} \tag{10}$$

where $\delta_m = -a(u - u_m, u_m(\Psi - 1))$.

It is obvious that $\delta_m \to 0$ at $m \to +\infty$. Integrating by parts (10), we have

$$\int_{Q_{\tau_2}} E(u_m)(\Psi - 1) dx\, dy\, dt \leq \int_{Q_{\tau_2}} P(u_m) \Psi' dx\, dy\, dt + \delta_m.$$

Hence

$$\int_{Q_{\tau_2}} E(u_m)(\Psi - 1) dx\, dy\, dt \leq \int_{Q_{\tau_2}\setminus Q_{\tau_1}} P(u_m) \mu \Psi dx\, dy\, dt + \delta_m. \tag{11}$$

The estimation (6) follows from (8) and (11) at $m \to +\infty$. □

Now we will estimate $\mu(y_1)$ in case when S_τ can be included to the $(n+m)$-dimensional parallelepiped which smallest edge is equal to $\lambda(\tau)$. Suppose that

$$\max_{S_\tau} \left\{ \left(\frac{1}{2}c^1 - c_{y_q}^{1q}\right), 0 \right\} = \gamma(\tau), \quad \max_{S_\tau} c_{p1} = \beta(\tau).$$

Applying the Friedreich and Cauchy–Bunyakovsky inequalities, we have from (9)

$$\left| \int_{S_\tau} P(v) dx\, dy'\, dt \right| \leq \left| \int_{S_\tau} c^{p1} v v_{y_p} dx\, dy'\, dt \right| + \left| \int_{S_\tau} \frac{1}{2}\left(c^1 - c_{y_q}^{1q}\right) v^2 dx\, dy'\, dt \right| \leq$$

$$\beta(\tau)\left[\int_{S_\tau} v^2 dx\, dy'\, dt\right]^{\frac{1}{2}}\left[\int_{S_\tau} v_{y_p}^2 dx\, dy'\, dt\right]^{\frac{1}{2}} + \gamma(\tau)\int_{S_\tau} v^2 dx\, dy'\, dt \leq$$

$$\left(\frac{\beta(\tau)\lambda(\tau)}{\pi\gamma_0} + \frac{\gamma(\tau)\lambda^2(\tau)}{\pi^2\gamma_0}\right)\int_{S_\tau} E(v) dx\, dy'\, dt.$$

Therefore we can set

$$\mu(\tau) = \pi^2\gamma_0\left(\pi\beta(\tau)\lambda(\tau) + \lambda^2(\tau)\gamma(\tau)\right)^{-1}.$$

If $\left(c^1 - 2c_{y_q}^{1q}\right) \leq 0$ in S_τ, then $\gamma(\tau) = 0$. Consequently

$$\mu(\tau) = \frac{\pi\gamma_0}{\beta(\tau)\lambda(\tau)}. \tag{12}$$

Example 1.

1. Let as $y_1 \geq \tau_1 \geq 0$, the domain Q lies inside the rotation body $|y'| \leq \frac{M}{2}(y_1+1)$, i.e., $\lambda(y_1) \leq M(y_1+1)$, $M > 0$. We have from (15)

$$\mu(y_1) = \frac{\pi c(y_1)}{M(y_1+1)}, \quad c(y_1) = \frac{d_0}{\beta(y_1)}.$$

Suppose that $c(x_1) = c = const > 0$.

In this case, from the inequality (6) we have

$$\int_{Q_{\tau_1}} E(u) dx\, dy\, dt \leq \Phi^{-1}(\tau_1,\tau_2)\int_{Q_{\tau_2}} E(u) dx\, dy\, dt \leq \left(\frac{\tau_1+1}{\tau_2+1}\right)^{\pi c}\int_{Q_{\tau_2}} E(u) dx\, dy\, dt.$$

2. Consider an example of Q for which

$$\lambda(y_1) \leq \pi c\left[(y_1+1)^{k-1}\right]^{-1}, k = const > 0.$$

It is clear that if $k > 1$, the domain Q is narrowing at $x_1 \to +\infty$. If $k = 1$, then $\lambda(x_1) \leq \pi c$ and this case includes domains lying in the band with the width πc. If $0 < k < 1$, then Q can be extended respectively at $x_1 \to +\infty$. For this kind of domains, we can assume

$$\mu(y_1) \leq (y_1+1)^{k-1}.$$

Then the estimate (6) is valid for considered domains if

$$\Phi^{-1}(\tau_1,\tau_2) = 2\exp\left[-(\tau_2+1)^k + (\tau_1+1)^k\right].$$

As a corollary of the Saint-Venant principle, we have the uniqueness theorem for the problem (1), (3) in unlimited domain Q for classes of functions increasing in infinity depending from $\lambda(\tau)$.

Theorem 2. Let $f(x,y,t) = 0$ in Q and conditions of theorem 1 hold. If $u(x,y,t)$ is a generalized solution of the problem (1), (3) in Q and for a sequence $\tau_m \to +\infty$ at $m \to +\infty$ and some $r_* = \text{const} > 0$,

$$\int_{Q_{\tau_m}} E(u)dx\,dy\,dt \leq \varepsilon(\tau_m)\Phi(r_*, \tau_m) \tag{13}$$

where $\varepsilon(\tau_m) \to 0$ at $\tau_m \to +\infty$, then $u = 0$ in Q_{r_*}.

Proof. We have from (6) considering (13)

$$\int_{Q_{r_*}} E(u)dx\,dy\,dt \leq \Phi^{-1}(r_*, \tau_m) \int_{Q_{\tau_2}} E(u)dx\,dy\,dt \leq \varepsilon(\tau_m) \to 0$$

at $\tau_m \to +\infty$. Hence $u = 0$ in Ω_{d_*}.

Further for any fixed $r_1 > r_*$, we have

$$\Phi(r_*, \tau_m) = e^{\int_{r_*}^{\tau_m} \mu(s)ds} = e^{\int_{r_1}^{\tau_m} \mu(s)ds} e^{\int_{r_*}^{r_1} \mu(s)ds} = c\Phi(r_1, \tau_m)$$

Therefore

$$\int_{Q_{e_1}} E(u)dx\,dy\,dt \leq \Phi^{-1}(r_1, \tau_m) \int_{Q_{\tau_m}} E(u)dx\,dy\,dt \leq \Phi^{-1}(r_1, \tau_m)\varepsilon(\tau_m)\Phi(r_*, \tau_m) =$$

$$c^{-1}\varepsilon(\tau_m) \to 0 \text{ as } \tau_m \to +\infty.$$

Hence, $u = 0$ in Q_{r_1}. Since r_1 was chosen arbitrary, $u = 0$ in Q. □

4. Conclusions

In the present paper, the analogy of the Saint-Venant principle is established for the generalized solution of the third order pseudoelliptical type equation. Furthermore, uniqueness theorems are obtained for solutions of the first boundary value problem in classes of functions increasing in infinity depending on the geometric characteristics of the domain $Q = D \times \Omega \times (0,T)$, were $D \subset \mathbb{R}_+^n = \{y : y_1 > 0\}$, Ω is bounded domain. Boundary value problems for the third order pseudoelliptical type equations in bounded domains were considered in [13].

The main goal of our research on these problems consists of the following parts:

(1) Establish energy estimates (analogous to the Saint-Venant's principle) that allow us to determine the widest class of uniqueness of solutions to the problem depending on the geometric characteristics of the domain.
(2) Construction of the solution of the problem under study on an unbounded domain in classes of functions growing at infinity.
(3) Establish estimates for solutions of the problem and its derivatives at infinitely remote boundary points.

The first part of our research on these problems is given in this paper. The remaining two parts will be studied in the future, which will be performed on the basis of this paper. Therefore, the results of this article are necessary and relevant for further qualitative research to solve third-order equations in the vicinity of irregular boundary points.

Author Contributions: Conceptualization, methodology, validation, formal analysis, investigation A.R.K.; validation, formal analysis, D.S. All authors have read and agreed to the published version of the manuscript.

Funding: The project is funded by the Institute of Technology and Business in České Budějovice, grant numbers: IGS 8210-004/2020 and IGS 8210-017/2020.

Conflicts of Interest: The authors declare no conflict of interest.

Abbreviations

The following abbreviations are used in this manuscript:

A.R.K   Abdukomil Risbekovich Khashimov
D.S.   Dana Smetanová

References

1. Barré de Saint-Venant, A.J.C. De la torsion des prismes. *Mem. Divers Savants Acad. Sci. Paris* **1855**, *14*, 233–560.
2. Gurtin, M.E. The Linear Theory of Elasticity. In *Handbuch der Physik*; Springee: Berlin, Germany. 1972.
3. Marin, M.; Oechsner, A.; Craciun, E.M. A generalization of the Saint-Venant's principle for an elastic body with dipolar structure. *Contin. Mech. Thermodyn.* **2020**, *32*, 269–278. [CrossRef]
4. Roohi, M.; Soleymani, K.; Salimi, M.; Heidari, M. Numerical evaluation of the general flow hydraulics and estimation of the river plain by solving the Saint-Venant equation. *Model. Earth Syst. Environ.* **2020**, *6*, 645–658. [CrossRef]
5. Xiao, B.; Sun, Z.X.; Shi, S.X.; Gao, C.; Lu, Q.C. On random elastic constraint conditions of Levinson beam model considering the violation of Saint-Venant's principle in dynamic. *Eur. Phys. J. Plus* **2020**, *135*, 168. [CrossRef]
6. Flavin, J.N. On Knowels version of Saint-Venant's principle two dimensional elastostatics. *Arch. Ration. Mech. Anal.* **1974**, *53*, 366–375. [CrossRef]
7. Knowels, J.K. On Saint-Venant's principle in the two-dimensional linear theory of elasticity. *Arch. Ration. Mech. Anal.* **1966**, *21*, 1–22. [CrossRef]
8. Oleinik, O.A.; Yosifian, G.A. On Saint-Venant's principle in the planar theory of elasticity. *Dokl. AN SSSR* **1978**, *239*, 530–533.
9. Worowich, I.I. Formulation of boundary value problems of the theory of elasticity at infinite energy integral and basis properties of homogeneous solutions. In *Mechanics of Deformable Bodies and Constructions*; Mashinistroenie: Moscow, Russia, 1975; pp. 112–118.
10. Galaktionov, V.A.; Shishkov, A.E. Saint-Venant's principle in blow-up for higher-order quasilinear parabolic equations. *Proc. Sec. A Math. R. Soc. Edinb.* **2003**, *133*, 1075–1119. [CrossRef]
11. Galaktionov, V.A.; Shishkov, A.E. Structure of boundary blow-up for higher order quasilinear parabolic PDE. *Proc. Roy. Soc. Lond. A* **2004**, *460*, 3299–3325. [CrossRef]
12. Khashimov, A.R. Estimation of derived any order of the solutions of the boundary value problems for equation of the third order of the composite type on infinity. *Uzb. Math. J.* **2016**, *2*, 140–148.
13. Kozhanov, A.I. *Boundary Value Problems for Equations of Mathematical Physics of the Odd Order*; Nauka: Novosibirsk, Russia, 1990; 149p.
14. Landis, E.M. On behavior of solutions of elliptic higher order equations in unlimited domains. *Trudy Mosk. Mat. Ob.* **1974**, *31*, 35–38.
15. Toupin, R. Saint-Venant's principle. *Arch. Rational Mech. Anal.* **1965**, *18*, 83–96. [CrossRef]
16. Fabiano, N.; Nikolić, N.; Shanmugam, T.; Radenović, S.; Čitaković, N. Tenth order boundary value problem solution existence by fixed point theorem. *J. Inequal. Appl.* **2020**, *166*, 1–11. [CrossRef]
17. Shanmugam, T.; Muthiah, M.; Radenović, S. Existence of Positive Solution for the Eighth-Order Boundary Value Problem Using Classical Version of Leray—Schauder Alternative Fixed Point Theorem. *Axioms* **2019**, *8*, 129. [CrossRef]
18. Todorčević, V. Subharmonic behavior and quasiconformal mappings. *Anal. Math. Phys.* **2019**, *9*, 1211–1225. [CrossRef]
19. Oleinik, O.A.; Radkevich, E.V. Analyticity and Liouville and Pharagmen–Lindelöf type theorems for general elliptic systems of differential equations. *Mate. Sb.* **1974**, *95*, 130–145.

20. Oleinik, O.A.; Yosifian, G.A. On singularities at the boundary points and uniqueness theorems for solutions of the first boundary value problem of elasticity. *Comm. Partial. Differ. Equ.* **1977**, *2*, 937–969. [CrossRef]
21. Oleinik, O.A.; Yosifian, G.A. Saint-Venant's principle for the mixed problem of the theory of elasticity and its applications. *Dokl. AN SSSR* **1977**, *233*, 824–827.
22. Berdichevsky, V.; Foster, D.J. On Saint-Venant's principle in the dynamics of elastic beams. *Int. J. Solids Struct.* **2003**, *40*, 3293–3310. [CrossRef]

© 2020 by the authors. Licensee MDPI, Basel, Switzerland. This article is an open access article distributed under the terms and conditions of the Creative Commons Attribution (CC BY) license (http://creativecommons.org/licenses/by/4.0/).

Article

Sufficient Conditions for Oscillation of Fourth-Order Neutral Differential Equations with Distributed Deviating Arguments

Omar Bazighifan [1,2,†], Feliz Minhos [3,*,†] and Osama Moaaz [4,†]

1. Department of Mathematics, Faculty of Science, Hadhramout University, Hadhramout 50512, Yemen; o.bazighifan@gmail.com
2. Department of Mathematics, Faculty of Education, Seiyun University, Hadhramout 50512, Yemen
3. Departamento de Matematica, Escola de Ciencias e Tecnologia, Centro de Investigacao em Matematica e Aplicacoes (CIMA), Instituto de Investigacao e Formacao Avancada, Universidade de Evora, Rua Romao Ramalho, 59, 7000-671 Evora, Portugal
4. Department of Mathematics, Faculty of Science, Mansoura University, Mansoura 35516, Egypt; o_moaaz@mans.edu.eg
* Correspondence: fminhos@uevora.pt
† These authors contributed equally to this work.

Received: 14 February 2020; Accepted: 8 April 2020; Published: 11 April 2020

Abstract: Some new sufficient conditions are established for the oscillation of fourth order neutral differential equations with continuously distributed delay of the form $\left(r(t)\left(N_x'''(t)\right)^\alpha\right)' + \int_a^b q(t,\vartheta)\, x^\beta\left(\delta(t,\vartheta)\right) d\vartheta = 0$, where $t \geq t_0$ and $N_x(t) := x(t) + p(t)\, x(\varphi(t))$. An example is provided to show the importance of these results.

Keywords: fourth-order differential equations; neutral delay; oscillation

1. Introduction

The theory of differential equations is an adequate mathematical apparatus for the simulation of processes and phenomena observed in biotechnology, neural networks, physics etc, see [1]. One area of active research in recent times is to study the sufficient criterion for oscillation of delay differential equations, see [1–28].

In this work, we establish the asymptotic behavior of fourth-order neutral differential equation of the form

$$\left(r(t)\left(N_x'''(t)\right)^\alpha\right)' + \int_a^b q(t,\vartheta)\, x^\beta\left(\delta(t,\vartheta)\right) d\vartheta = 0, \tag{1}$$

where $t \geq t_0$ and $N_x(t) := x(t) + p(t)\, x(\varphi(t))$. In this paper, we assume that:

A1: α and β are a quotient of odd positive integers and $\beta \geq \alpha$;
A2: $r, p \in C[t_0, \infty)$, $r(t) > 0$, $r'(t) \geq 0$ and $\int^\infty r^{-1/\alpha}(s)\, ds = \infty$;
A3: $q \in C\left([t_0, \infty) \times (a,b), \mathbb{R}\right)$, $q(t,\vartheta) > 0$, $0 \leq p(t) < p_0 < \infty$ and $q(t)$ is not identically zero for large t;
A4: $\varphi \in C^1[t_0, \infty)$, $\delta \in p\left([t_0, \infty) \times (a,b), \mathbb{R}\right)$, $\varphi'(t) > 0$, $\varphi(t) \leq t$, $\lim_{t \to \infty} \varphi(t) = \lim_{t \to \infty} \delta(t,\vartheta) = \infty$ and $\delta(t,\vartheta)$ has nondecreasing.

Definition 1. *The function $x \in C^3[t_y, \infty)$, $t_y \geq t_0$, is called a solution of (1), if $r(t)\left(N_x'''(t)\right)^\alpha \in C^1[t_y, \infty)$, and $x(t)$ satisfies (1) on $[t_y, \infty)$.*

Definition 2. *A solution of (1) is called oscillatory if it has arbitrarily large zeros on $[t_x, \infty)$, and otherwise is called to be nonoscillatory.*

Definition 3. *The Equation (1) is called oscillatory if every its solutions are oscillatory.*

In the following, we discuss some important papers:

Chatzarakis et al. [9] proved the equation (1) where $\alpha = \beta$, is oscillatory, if

$$\int_{t_0}^{\infty} \left(\varpi(s) - \frac{2^\alpha r(s)}{\mu^\alpha s^{2\alpha} \rho^\alpha(s)} \left(\frac{\rho'(s)}{\alpha+1} \right)^{\alpha+1} \right) ds = \infty,$$

for some $\mu \in (0,1)$ and

$$\int_{t_0}^{\infty} \left(\vartheta(s) \left(\int_t^\infty (Q(v))^{\frac{1}{\alpha}} r^{\frac{-1}{\alpha}}(v) \, dv \right) - \frac{\theta'^2_+(s)}{4\theta(s)} \right) ds = \infty,$$

where $\varpi(t) := k\rho(t) Q(t) (1 - p(\delta(t,a)))^\alpha (\delta(t,a) \setminus t)^{3\alpha}$ and $\rho, \theta \in C^1([v_0, \infty), (0, \infty))$.

Moaaz et al. in [19] extended the Riccati transformation to obtain new oscillatory criteria for (1) as condition

$$\int_{t_0}^{\infty} \left[\theta(s) Q(s) - \frac{1}{\lambda 4} \left(\frac{\theta'(s)}{\theta(s)} \right)^2 \right] ds = \infty,$$

where $\lambda \in (0,1)$ and a function $\theta \in C^1([v_0, \infty), (0, \infty))$.

Authors in [24] studied oscillatory behavior of equation

$$N_x^{(n)}(t) + q(t) x(\delta(t)) = 0, \tag{2}$$

where n is even, they proved it oscillatory by using the Riccati transformation if either

$$\liminf_{t \to \infty} \int_{\varphi(t)}^t Q(s) \, ds > \frac{(n-1)!}{e}, \tag{3}$$

or

$$\limsup_{t \to \infty} \int_{\varphi(t)}^t Q(s) \, ds > (n-1)!,$$

where $Q(t) := \varphi^{n-1}(t) (1 - p(\varphi(t))) q(t)$.

Xing et al. [22] proved that the even-order differential equation

$$\left(r(t) \left(N_x^{(n-1)}(t) \right)^\alpha \right)' + q(t) x^\beta(\delta(t)) = 0,$$

is oscillatory, if

$$\left(\delta^{-1}(t) \right)' \geq \delta_0 > 0, \ \varphi'(t) \geq \varphi_0 > 0, \ \varphi^{-1}(\delta(t)) < t$$

and

$$\liminf_{t \to \infty} \int_{\varphi^{-1}(\delta(t))}^t \frac{\widehat{q}(s)}{r(s)} \left(s^{n-1} \right)^\alpha ds > \frac{\left(\frac{1}{\delta_0} + \frac{p_0^\alpha}{\delta_0 \varphi_0} \right)}{e \left((n-1)! \right)^{-\alpha}}, \tag{4}$$

where $\widehat{q}(t) := \min \{ q(\delta^{-1}(t)), q(\delta^{-1}(\varphi(t))) \}$ and n is even.

To prove this, we apply the previous results to the equation

$$(x(t) + px(\varphi t))^{(n)} + bx(\delta t) = 0, \ t \geq 1, \tag{5}$$

where $n = 4$, $p = 7/8$, $\varphi = 1/e$, $\delta = 1/e^2$ and $b = q_0/v^4$, we find:

1. By applying condition (3) in (5), we find

$$q_0 > 3561.9.$$

2. By applying condition (4) in (5), we get

$$q_0 > 3008.5.$$

Hence, [22] improved the results in [24].

Thus, the motivation in studying this paper is complement results in [9] and improve results [22,24].

By using the Riccati transformations, we establish a new oscillation criterion for a class of fourth-order neutral differential equations (1). An example is provided to illustrate the main results.

2. Some Auxiliary Lemmas

We shall employ the following lemmas

Lemma 1 ([3]). *Let $x \in C^n([t_0, \infty), (0, \infty))$. Assume that $x^{(n)}(t)$ is of fixed sign and not identically zero on $[t_0, \infty)$ and there exists a $t_1 \geq t_0$ such that $x^{(n-1)}(t) x^{(n)}(t) \leq 0$ for all $t \geq t_1$. If $\lim_{t \to \infty} x(t) \neq 0$, then for every $\mu \in (0,1)$ there exists $t_\mu \geq t_1$ such that*

$$x(t) \geq \frac{\mu}{(n-1)!} t^{n-1} \left| x^{(n-1)}(t) \right| \text{ for } t \geq t_\mu.$$

Lemma 2 ([16]). *Let the function x satisfies $x^{(i)}(t) > 0$, $i = 0, 1, ..., n$, and $x^{(n+1)}(t) < 0$, then*

$$\frac{x(t)}{t^n/n!} \geq \frac{x'(t)}{t^{n-1}/(n-1)!}.$$

Lemma 3 ([4]). *Assume that $x, v \geq 0$ and $\alpha \geq 1$ is a positive real number. Then*

$$(x+v)^\alpha \leq 2^{\alpha-1}(x^\alpha + v^\alpha)$$

and

$$(x+v)^\beta \leq x^\beta + v^\beta, \text{ for } \beta \leq 1.$$

Lemma 4 ([9]). *Assume that x is an eventually positive solution of (1). Then, there exist two possible cases:*

(S_1) $N_x^{(\kappa)}(t) > 0$ for $\kappa = 0, 1, 2, 3;$
(S_2) $N_x(t) > 0$, $N_x'(t) > 0$, $N_x''(t) < 0$ and $N_x'''(t) > 0,$

for $t \geq t_1$, where $t_1 \geq t_0$ is sufficiently large.

Notation 1. *We consider the following notations:*

$$p_1(t) = \frac{1}{p(\varphi^{-1}(t))}\left(1 - \frac{(\varphi^{-1}(\varphi^{-1}(t)))^3}{(\varphi^{-1}(t))^3 p(\varphi^{-1}(\varphi^{-1}(t)))}\right),$$

$$p_2(t) = \frac{1}{p(\varphi^{-1}(t))}\left(1 - \frac{(\varphi^{-1}(\varphi^{-1}(t)))}{(\varphi^{-1}(t)) p(\varphi^{-1}(\varphi^{-1}(t)))}\right)$$

$$\Psi(t) = M_1^{\beta-\alpha} \theta(t) \int_a^b q(t,\vartheta) p_1^\beta(\delta(t,\vartheta)) \, d\vartheta$$

$$\tilde{R}(t) = \int_a^b \left(\frac{\mu(\varphi^{-1}(\eta(t,\vartheta)))^3}{6}\right)^\beta q(t,\vartheta) p_1^\beta(\eta(t,\vartheta)) r^{-\beta/\alpha}\left(\varphi^{-1}(\eta(t,\vartheta))\right) d\vartheta$$

$$R(t) = \int_t^\infty \left(\frac{1}{r(\varrho)} \int_\varrho^\infty \left(\int_a^b q(s,\vartheta) \left(\frac{\varphi^{-1}(\sigma(s,\vartheta))}{s}\right)^\beta d\vartheta\right) ds\right)^{1/\alpha} d\varrho,$$

and

$$\Phi(t) := p_2^{\beta/\alpha} \theta_1(t) M_2^{(\beta-\alpha)/\alpha} \int_t^\infty \left(\frac{1}{r(\varrho)} \int_\varrho^\infty \left(\int_a^b q(s,\vartheta) \left(\frac{\varphi^{-1}(\delta(s,\vartheta))}{s}\right)^\beta d\vartheta\right) ds\right)^{1/\alpha} d\varrho.$$

3. Main Results

In this part, we will discuss some oscillation criteria for Equation (1).

Lemma 5. *Assume that x is an eventually positive solution of (1) and*

$$\left(\varphi^{-1}\left(\varphi^{-1}(t)\right)\right)^3 < \left(\varphi^{-1}(t)\right)^3 p\left(\varphi^{-1}\left(\varphi^{-1}(t)\right)\right). \tag{6}$$

Then

$$x(t) \geq \frac{1}{p(\varphi^{-1}(t))}\left(N_x\left(\varphi^{-1}(t)\right) - \frac{1}{p(\varphi^{-1}(\varphi^{-1}(t)))} N_x\left(\varphi^{-1}\left(\varphi^{-1}(t)\right)\right)\right). \tag{7}$$

Proof. Let x be an eventually positive solution of (1) on $[t_0, \infty)$. From the definition of $z(t)$, we see that

$$p(t) x(\varphi(t)) = N_x(t) - x(t),$$

and so

$$p\left(\varphi^{-1}(t)\right) x(t) = N_x\left(\varphi^{-1}(t)\right) - x\left(\varphi^{-1}(t)\right).$$

Repeating the same process, we obtain

$$x(t) = \frac{1}{p(\varphi^{-1}(t))}\left(N_x\left(\varphi^{-1}(t)\right) - \left(\frac{N_x(\varphi^{-1}(\varphi^{-1}(t)))}{p(\varphi^{-1}(\varphi^{-1}(t)))} - \frac{x(\varphi^{-1}(\varphi^{-1}(t)))}{p(\varphi^{-1}(\varphi^{-1}(t)))}\right)\right),$$

which yields

$$x(t) \geq \frac{N_x(\varphi^{-1}(t))}{p(\varphi^{-1}(t))} - \frac{1}{p(\varphi^{-1}(t))} \frac{N_x(\varphi^{-1}(\varphi^{-1}(t)))}{p(\varphi^{-1}(\varphi^{-1}(t)))}.$$

Thus, (7) holds. This completes the proof. □

Theorem 1. Let $\delta(t) \leq \varphi(t)$ and (6) holds. If there exist positive functions $\theta, \theta_1 \in C^1([t_0, \infty), \mathbb{R})$ such that

$$\int_{t_0}^{\infty} \left(\Psi(s) - \frac{2^\alpha}{(\alpha+1)^{\alpha+1}} \frac{r(\varphi^{-1}(\delta(s,a))) (\theta'(s))^{\alpha+1}}{\left(\mu_1 \theta(s) (\varphi^{-1}(\delta(s,a)))' (\delta(s,a))' (\varphi^{-1}(\delta(s,a)))^2 \right)^\alpha} \right) ds = \infty \quad (8)$$

and

$$\int_{t_0}^{\infty} \left(\Phi(s) - \frac{(\theta_1'(s))^2}{4\theta_1(s)} \right) ds = \infty, \quad (9)$$

for some $\mu_1 \in (0,1)$ and every $M_1, M_2 > 0$, then (1) is oscillatory.

Proof. Let x be a non-oscillatory solution of (1) on $[t_0, \infty)$. Without loss of generality, we can assume that x is eventually positive. It follows from Lemma 4 that there exist two possible cases (S_1) and (S_2).

Let (S_1) holds. From Lemma 2, we obtain $N_x(t) \geq \frac{1}{3} t N_x'(t)$ and hence the function $t^{-3} N_x(t)$ is nonincreasing, which with the fact that $\varphi^{-1}(t) \leq \varphi^{-1}(\varphi^{-1}(t))$ gives

$$\left(\varphi^{-1}(t) \right)^3 N_x\left(\varphi^{-1}\left(\varphi^{-1}(t) \right) \right) \leq \left(\varphi^{-1}\left(\varphi^{-1}(t) \right) \right)^3 N_x\left(\varphi^{-1}(t) \right). \quad (10)$$

From (7) and (10), we get that

$$\begin{aligned}
x(t) &\geq \frac{N_x(\varphi^{-1}(t))}{p(\varphi^{-1}(t))} \left(1 - \frac{(\varphi^{-1}(\varphi^{-1}(t)))^{n-1}}{(\varphi^{-1}(t))^{n-1} p(\varphi^{-1}(\varphi^{-1}(t)))} \right) \\
&\geq p_1(t) N_x\left(\varphi^{-1}(t) \right). \quad (11)
\end{aligned}$$

From (1) and (11), we obtain

$$\left(r(t) (N_x'''(t))^\alpha \right)' + \int_a^b q(t, \vartheta) p_1^\beta(\delta(t, \vartheta)) N_x^\beta\left(\varphi^{-1}(\delta(t, \vartheta)) \right) d\vartheta \leq 0. \quad (12)$$

Since $\delta(t, \xi)$ is nondecreasing with respect to s, we get $\delta(t, \vartheta) \geq \delta(t, a)$ for $\xi \in (a, b)$ and so

$$\left(r(t) (N_x'''(t))^\alpha \right)' + N_x^\beta\left(\varphi^{-1}(\delta(t, a)) \right) \int_a^b q(t, \vartheta) p_1^\beta(\delta(t, \vartheta)) d\vartheta \leq 0.$$

Next, we define a function ω by

$$\omega(t) := \theta(t) \frac{r(t) (N_x'''(t))^\alpha}{N_x^\alpha(\varphi^{-1}(\delta(t, a)))} > 0.$$

Differentiating and using (12), we obtain

$$\begin{aligned}
\omega'(t) \leq{}& \frac{\theta'(t)}{\theta(t)} \omega(t) - \theta(t) N_x^{\beta-\alpha}\left(\varphi^{-1}(\delta(t, a)) \right) \int_a^b q(t, \vartheta) p_1^\beta(\delta(t, \vartheta)) d\vartheta \\
& - \alpha \theta(t) \frac{r(t) (N_x'''(t))^\alpha \left(\varphi^{-1}(\delta(t, a)) \right)' (\delta(t, a))' N_x'\left(\varphi^{-1}(\delta(t, a)) \right)}{N_x^{\alpha+1}(\varphi^{-1}(\delta(t, a)))}. \quad (13)
\end{aligned}$$

Recalling that $r(t) (N_x'''(t))^\alpha$ is decreasing, we get

$$r\left(\varphi^{-1}(\delta(t, a)) \right) \left(N_x'''\left(\varphi^{-1}(\delta(t, a)) \right) \right)^\alpha \geq r(t) (N_x'''(t))^\alpha.$$

This yields

$$\left(N_x''' \left(\varphi^{-1}\left(\delta\left(t,a\right)\right)\right)\right)^\alpha \geq \frac{r(t)}{r\left(\varphi^{-1}\left(\delta\left(t,a\right)\right)\right)} \left(N_x'''(t)\right)^\alpha. \tag{14}$$

It follows from Lemma 1 that

$$N_x'\left(\varphi^{-1}\left(\delta\left(t,a\right)\right)\right) \geq \frac{\mu_1}{2}\left(\varphi^{-1}\left(\delta\left(t,a\right)\right)\right)^2 N_x'''\left(\varphi^{-1}\left(\delta\left(t,a\right)\right)\right), \tag{15}$$

for all $\mu_1 \in (0,1)$. Thus, by (13)–(15), we get

$$\omega'(t) \leq \frac{\theta'(t)}{\theta(t)}\omega(t) - \theta(t) N_x^{\beta-\alpha}\left(\varphi^{-1}\left(\delta\left(t,a\right)\right)\right) \int_a^b q(t,\vartheta) p_1^\beta(\delta(t,\vartheta)) d\vartheta$$
$$-\alpha\theta(t)\frac{\mu_1}{2}\left(\frac{r(t)}{r\left(\varphi^{-1}\left(\delta\left(t,a\right)\right)\right)}\right)^{1/\alpha} \frac{r(t)\left(N_x'''(t)\right)^{\alpha+1}\left(\varphi^{-1}\left(\delta\left(t,a\right)\right)\right)'\left(\delta\left(t,a\right)\right)'\left(\varphi^{-1}\left(\delta\left(t,a\right)\right)\right)^2}{N_x^{\alpha+1}\left(\varphi^{-1}\left(\delta\left(t,a\right)\right)\right)}$$

Hence,

$$\omega'(t) \leq \frac{\theta'(t)}{\theta(t)}\omega(t) - \theta(t) N_x^{\beta-\alpha}\left(\varphi^{-1}\left(\delta\left(t,a\right)\right)\right) \int_a^b q(t,\vartheta) p_1^\beta(\delta(t,\vartheta)) d\vartheta$$
$$-\alpha\frac{\mu_1}{2}\left(\frac{r(t)}{r\left(\varphi^{-1}\left(\delta\left(t,a\right)\right)\right)}\right)^{1/\alpha} \frac{\left(\varphi^{-1}\left(\delta\left(t,a\right)\right)\right)'\left(\delta\left(t,a\right)\right)'\left(\varphi^{-1}\left(\delta\left(t,a\right)\right)\right)^2}{(r\theta)^{1/\alpha}(t)} \omega^{\frac{\alpha+1}{\alpha}}(t).$$

Since $N_x'(t) > 0$, there exist a $t_2 \geq t_1$ and a constant $M > 0$ such that

$$N_x(t) > M, \tag{16}$$

for all $t \geq t_2$. Using the inequality

$$Ux - Vx^{(\beta+1)/\beta} \leq \frac{\beta^\beta}{(\beta+1)^{\beta+1}} \frac{U^{\beta+1}}{V^\beta}, \quad V > 0,$$

with

$$U = \frac{\theta'(t)}{\theta(t)}, \quad V = \alpha\frac{\mu_1}{2}\left(\frac{r(t)}{r\left(\varphi^{-1}\left(\delta\left(t,a\right)\right)\right)}\right)^{1/\alpha} \frac{\left(\varphi^{-1}\left(\delta\left(t,a\right)\right)\right)'\left(\delta\left(t,a\right)\right)'\left(\varphi^{-1}\left(\delta\left(t,a\right)\right)\right)^2}{(r\theta)^{1/\alpha}(t)}$$

and $x = \omega$, we get

$$\omega'(t) \leq -\Psi(t) + \frac{2^\alpha}{(\alpha+1)^{\alpha+1}} \frac{r\left(\varphi^{-1}\left(\delta\left(t,a\right)\right)\right)\left(\theta'(t)\right)^{\alpha+1}}{\left(\mu_1\theta(t)\left(\varphi^{-1}\left(\delta\left(t,a\right)\right)\right)'\left(\delta\left(t,a\right)\right)'\left(\varphi^{-1}\left(\delta\left(t,a\right)\right)\right)^2\right)^\alpha}.$$

This implies that

$$\int_{t_1}^t \left(\Psi(s) - \frac{2^\alpha}{(\alpha+1)^{\alpha+1}} \frac{r\left(\varphi^{-1}\left(\delta\left(t,a\right)\right)\right)\left(\theta'(t)\right)^{\alpha+1}}{\left(\mu_1\theta(t)\left(\varphi^{-1}\left(\delta\left(t,a\right)\right)\right)'\left(\delta\left(t,a\right)\right)'\left(\varphi^{-1}\left(\delta\left(t,a\right)\right)\right)^2\right)^\alpha}\right) ds \leq \omega(t_1),$$

which contradicts (8).

In the case where (S_2) satisfies, by using Lemma 2, we find that

$$N_x(t) \geq t N_x'(t) \tag{17}$$

and hence $\left(t^{-1} N_x(t)\right)' \leq 0$. Therefore,

$$\varphi^{-1}(t) N_x\left(\varphi^{-1}\left(\varphi^{-1}(t)\right)\right) \leq \varphi^{-1}\left(\varphi^{-1}(t)\right) N_x\left(\varphi^{-1}(t)\right). \tag{18}$$

From (7) and (18), we have

$$\begin{aligned} x(t) &\geq \frac{1}{p\left(\varphi^{-1}(t)\right)} \left(1 - \frac{\left(\varphi^{-1}\left(\varphi^{-1}(t)\right)\right)}{\left(\varphi^{-1}(t)\right) p\left(\varphi^{-1}\left(\varphi^{-1}(t)\right)\right)}\right) N_x\left(\varphi^{-1}(t)\right) \\ &= p_2(t) N_x\left(\varphi^{-1}(t)\right), \end{aligned}$$

which with (1) gives

$$\left(r(t) \left(N_x'''(t)\right)^{\alpha}\right)' \leq -\int_a^b q(t,\vartheta) p_2^{\beta}(\delta(t,\vartheta)) N_x^{\beta}\left(\varphi^{-1}(\delta(t,\vartheta))\right) d\vartheta.$$

Integrating this inequality from t to ϱ, we obtain

$$r(\varrho)\left(N_x'''(\varrho)\right)^{\alpha} - r(t)\left(N_x'''(t)\right)^{\alpha} \leq -\int_t^{\varrho}\left(\int_a^b q(t,\vartheta) p_2^{\beta}(\delta(t,\vartheta)) N_x^{\beta}\left(\varphi^{-1}(\delta(t,\vartheta))\right) d\vartheta\right) ds. \tag{19}$$

From (17), we get that

$$N_x\left(\varphi^{-1}(\delta(t,\vartheta))\right) \geq \frac{\varphi^{-1}(\delta(t,\vartheta))}{t} N_x(t). \tag{20}$$

Letting $\varrho \to \infty$ in (19) and using (20), we obtain

$$r(t)\left(N_x'''(t)\right)^{\alpha} \geq p_2^{\beta}(\delta(t,a)) N_x^{\beta}(t) \int_t^{\infty}\left(\int_a^b q(s,\vartheta) \left(\frac{\varphi^{-1}(\delta(s,\vartheta))}{s}\right)^{\beta} d\vartheta\right) ds.$$

Integrating this inequality again from t to ∞, we get

$$N_x''(t) \leq -p_2^{\beta/\alpha} N_x^{\beta/\alpha}(t) \int_t^{\infty} \left(\frac{1}{r(\varrho)} \int_{\varrho}^{\infty}\left(\int_a^b q(s,\vartheta)\left(\frac{\varphi^{-1}(\delta(s,\vartheta))}{s}\right)^{\beta} d\vartheta\right) ds\right)^{1/\alpha} d\varrho, \tag{21}$$

for all $\mu_2 \in (0,1)$.

Now, we define

$$w(t) = \theta_1(t) \frac{N_x'(t)}{N_x(t)}.$$

Then $w(t) > 0$ for $t \geq t_1$. By differentiating w and using (21), we find

$$\begin{aligned} w'(t) &= \frac{\theta_1'(t)}{\theta_1(t)} w(t) + \theta_1(t) \frac{N_x''(t)}{N_x(t)} - \theta_1(t)\left(\frac{N_x'(t)}{N_x(t)}\right)^2 \\ &\leq \frac{\theta_1'(t)}{\theta_1(t)} w(t) - \frac{1}{\theta_1(t)} w^2(t) \\ &\quad - p_2^{\beta/\alpha} \theta_1(t) N_x^{\beta/\alpha - 1}(t) \int_t^{\infty}\left(\frac{1}{r(\varrho)} \int_{\varrho}^{\infty}\left(\int_a^b q(s,\vartheta)\left(\frac{\varphi^{-1}(\delta(s,\vartheta))}{s}\right)^{\beta} d\vartheta\right) ds\right)^{1/\alpha} d\varrho. \end{aligned}$$

Thus, we obtain

$$w'(t) \leq -\Phi(t) + \frac{\theta_1'(t)}{\theta_1(t)} w(t) - \frac{1}{\theta_1(t)} w^2(t),$$

and so
$$w'(t) \leq -\Phi(t) + \frac{(\theta_1'(t))^2}{4\theta_1(t)}.$$

Then, we get
$$\int_{t_1}^{t} \left(\Phi(s) - \frac{(\theta'(t))^2}{4\theta(t)} \right) ds \leq w(t_1),$$

which contradicts (9). This completes the proof. □

Theorem 2. *Let*
$$\frac{\left(\varphi^{-1}\left(\varphi^{-1}(t)\right)\right)^{n-1}}{\left(\varphi^{-1}(t)\right)^{n-1} p\left(\varphi^{-1}\left(\varphi^{-1}(t)\right)\right)} \leq 1. \tag{22}$$

Suppose that there exist positive functions $\eta, \sigma \in p^1([t_0, \infty), \mathbb{R})$ *satisfying*

$$\eta(t) \leq \delta(t), \ \eta(t) < \varphi(t), \ \sigma(t) \leq \delta(t), \ \sigma(t) < \varphi(t), \ \sigma'(t) \geq 0 \text{ and } \lim_{t \to \infty} \eta(t) = \lim_{t \to \infty} \sigma(t) = \infty. \tag{23}$$

If the equations
$$\psi'(t) + \tilde{R}(t) \psi^{\beta/\alpha} \left(\varphi^{-1}(\eta(t,a)) \right) = 0 \tag{24}$$

and
$$\phi'(t) + p_2^{\beta/\alpha} \left(\varphi^{-1}(\sigma(t,a)) \right)^{\beta/\alpha} R(t) \phi^{\beta/\alpha} \left(\varphi^{-1}(\sigma(t,a)) \right) = 0 \tag{25}$$

are oscillatory, then (1) is oscillatory.

Proof. Let x be a non-oscillatory solution of (1) on $[t_0, \infty)$. Without loss of generality, we suppose that $x > 0$. From Lemma 4, we find there exist two possible cases (S_1) and (S_2).

Assume that Case (S_1) holds. From Theorem 1, we get that (12) holds. Since $\eta(t) \leq \delta(t)$ and $z'(t) > 0$, we obtain

$$\left(r(t) \left(N_x'''(t) \right)^{\alpha} \right)' \leq -\int_a^b q(t,\vartheta) \, p_1^{\beta}(\eta(t,\vartheta)) \, N_x^{\beta} \left(\varphi^{-1}(\eta(t,\vartheta)) \right) d\vartheta. \tag{26}$$

Now, by using Lemma 1, we have
$$N_x(t) \geq \frac{\mu}{6} t^3 N_x'''(t). \tag{27}$$

for some $\mu \in (0,1)$. It follows from (26) and (27) that, for all $\mu \in (0,1)$,

$$\left(r(t) \left(N_x'''(t) \right)^{\alpha} \right)' + \int_a^b \left(\frac{\mu \left(\varphi^{-1}(\eta(t,\vartheta)) \right)^3}{6} \right)^{\beta} q(t,\vartheta) \, p_1^{\beta}(\eta(t,\vartheta)) \left(N_x''' \left(\varphi^{-1}(\eta(t,\vartheta)) \right) \right)^{\beta} d\vartheta \leq 0.$$

Thus, we choose
$$\psi(t) = r(t) \left(N_x'''(t) \right)^{\alpha}.$$

So, we find that ψ is a positive solution of the inequality
$$\psi'(t) + \tilde{R}(t) \psi^{\beta/\alpha} \left(\varphi^{-1}(\eta(t,a)) \right) \leq 0.$$

Using (see ([15] Theorem 1)), we see (24) also has a positive solution, a contradiction.

Suppose that Case (S_2) holds. From Theorem 1, we get that (21) holds. Since $\sigma(t) \leq \delta(t)$ and $N_x'(t) > 0$, we have that

$$N_x''(t) \leq -p_2^{\beta/\alpha} N_x^{\beta/\alpha} \left(\varphi^{-1}(\sigma(t,a))\right) \int_t^\infty \left(\frac{1}{r(\varrho)} \int_\varrho^\infty \left(\int_a^b q(s,\vartheta) \left(\frac{\varphi^{-1}(\sigma(s,\vartheta))}{s}\right)^\beta d\vartheta\right) ds\right)^{1/\alpha} d\varrho, \quad (28)$$

Using Lemma 2, we get that

$$N_x(t) \geq t N_x'(t). \quad (29)$$

From (18) and (29), we obtain

$$N_x''(t) \leq -p_2^{\beta/\alpha} \left(N_x'\left(\varphi^{-1}(\sigma(t,a))\right)\right)^{\beta/\alpha} \left(\varphi^{-1}(\sigma(t,a))\right)^{\beta/\alpha} R(t).$$

Now, we choose $\phi(t) := N_x'(t)$, thus, we find that ϕ is a positive solution of

$$\phi'(t) + p_2^{\beta/\alpha} \left(\varphi^{-1}(\sigma(t,a))\right)^{\beta/\alpha} R(t) \phi^{\beta/\alpha}\left(\varphi^{-1}(\sigma(t,a))\right) \leq 0. \quad (30)$$

Using (see ([15] Theorem 1)), we see (25) also has a positive solution, a contradiction. The proof is complete. □

Example 1. *Consider the differential equation*

$$\left(\left[x(t) + \frac{1}{2}x\left(\frac{t}{3}\right)\right]'''\right)' + \int_0^1 \left(\frac{q_0}{t^4}\right) \vartheta x \left(\frac{t-\xi}{2}\right) d\vartheta = 0, \quad (31)$$

where $q_0 > 0$ is a constant. Let $\alpha = \beta = 1$, $r(t) = 1$, $p(t) = 1/2$, $\varphi(t) = t/3$, $\varphi^{-1}(t) = 3t$, $\xi(t,a) = t/2$, $q(t,\vartheta) = (q_0 \backslash t^4) \vartheta$.

Thus, by using Theorem 1, then Equation (31) is oscillatory.

Remark 1. *By applying our results in (5), we see that our results improve [22,24].*

Remark 2. *One can easily see that the results obtained in [24] cannot be applied to conditions in Theorem 1, so our results are new.*

4. Conclusions

In this work, our method is based on using the Riccati transformations to get some oscillation criteria of (1). There are numerous results concerning the oscillation criteria of fourth order equations, which include various forms of criteria as Hille/Nehari, Philos, etc. This allows us to obtain also various criteria for the oscillation of (1). Further, we can try to get some oscillation criteria of (1) if $N_x(t) := x(t) - p(t) x(\varphi(t))$ in the future work.

Author Contributions: The authors claim to have contributed equally and significantly in this paper. All authors have read and agreed to the published version of the manuscript.

Funding: The authors received no direct funding for this work.

Acknowledgments: The authors thank the reviewers for for their useful comments, which led to the improvement of the content of the paper.

Conflicts of Interest: There are no competing interests between the authors.

References

1. Hale, J.K. *Theory of Functional Differential Equations*; Springer: New York, NY, USA, 1977.
2. Agarwal, R.P.; Bohner, M.; Li, T.; Zhang, C. A new approach in the study of oscillatory behavior of even-order neutral delay diferential equations. *Appl. Math. Comput.* **2013**, *225*, 787–794.
3. Agarwal, R.; Grace, S.; O'Regan, D. *Oscillation Theory for Difference and Functional Differential Equations*; Kluwer Acad. Publ.: Dordrecht, The Netherlands, 2000.
4. Baculikova, B.; Dzurina, J. Oscillation theorems for second-order nonlinear neutral differential equations. *Comput. Math. Appl.* **2011**, *62*, 4472–4478. [CrossRef]
5. Bazighifan, O.; Cesarano, C. Some New Oscillation Criteria for Second-Order Neutral Differential Equations with Delayed Arguments. *Mathematics* **2019**, *7*, 619. [CrossRef]
6. Bazighifan, O.; Elabbasy, M.E.; Moaaz, O. Oscillation of higher-order differential equations with distributed delay. *J. Inequal. Appl.* **2019**, *55*, 1–9. [CrossRef]
7. Bazighifan, O.; Postolache, M. An improved conditions for oscillation of functional nonlinear differential equations. *Mathematics* **2020**, *8*, 552. [CrossRef]
8. Bazighifan, O. An Approach for Studying Asymptotic Properties of Solutions of Neutral Differential Equations. *Symmetry* **2020**, *12*, 555. [CrossRef]
9. Chatzarakis, G.E.; Elabbasy, E.M.; Bazighifan, O. An oscillation criterion in 4th-order neutral differential equations with a continuously distributed delay. *Adv. Differ. Equ.* **2019**, *336*, 1–9.
10. Chatzarakis, G.E.; Li, T. Oscillation criteria for delay and advanced differential equations with nonmonotone arguments. *Complexity* **2018**, *2018*, 8237634. [CrossRef]
11. El-Nabulsi, R.A.; Moaaz, O.; Bazighifan, O. New Results for Oscillatory Behavior of Fourth-Order Differential Equations. *Symmetry* **2020**, *12*, 136. [CrossRef]
12. Elabbasy, E.M.; Cesarano, C.; Bazighifan, O.; Moaaz, O. Asymptotic and oscillatory behavior of solutions of a class of higher order differential equation. *Symmetry* **2019**, *11*, 1434. [CrossRef]
13. Elabbasy, E.M.; Hassan, T.S.; Moaaz, O. Oscillation behavior of second-order nonlinear neutral differential equations with deviating arguments. *Opusc. Math.* **2012**, *32*, 719–730. [CrossRef]
14. Li, T.; Han, Z.; Zhao, P.; Sun, S. Oscillation of even-order neutral delay differential equations. *Adv. Differ. Equ.* **2010**, *127*, 503–509.
15. Philos, C.G. On the existence of non-oscillatory solutions tending to zero at ∞ for differential equations with positive delays. *Arch. Math.* **1981**, *36*, 168–178. [CrossRef]
16. Kiguradze, I.T.; Chanturiya, T.A. *Asymptotic Properties of Solutions of Nonautonomous Ordinary Differential Equations*; Kluwer Acad. Publ.: Dordrecht, The Netherlands, 1993.
17. Moaaz, O.; Dassios, I.; Bazighifan, O.; Muhib, A. Oscillation Theorems for Nonlinear Differential Equations of Fourth-Order. *Mathematics* **2020**, *8*, 520. [CrossRef]
18. Moaaz, O.; Elabbasy, E.M.; Bazighifan, O. On the asymptotic behavior of fourth-order functional differential equations. *Adv. Differ. Equ.* **2017**, *2017*, 261. [CrossRef]
19. Moaaz, O.; Elabbasy, E.M.; Muhib, A. Oscillation criteria for even-order neutral differential equations with distributed deviating arguments. *Adv. Differ. Equ.* **2019**, *297*, 1–10. [CrossRef]
20. Moaaz, O.; Kumam, P.; Bazighifan, O. On the Oscillatory Behavior of a Class of Fourth-Order Nonlinear Differential Equation. *Symmetry* **2020**, *12*, 524. [CrossRef]
21. Minhos, F.; de Sousa, R. Solvability of Coupled Systems of Generalized Hammerstein-Type Integral Equations in the Real Line. *Mathematics* **2019**, *8*, 111. [CrossRef]
22. Xing, G.; Li, T.; Zhang, C. Oscillation of higher-order quasi linear neutral differential equations. *Adv. Differ. Equ.* **2011**, *2011*, 45. [CrossRef]
23. Zafer, A. Oscillation criteria for even order neutral differential equations. *Appl. Math. Lett.* **1998**, *11*, 21–25. [CrossRef]
24. Zhang, Q.; Yan, J. Oscillation behavior of even order neutral differential equations with variable coefficients. *Appl. Math. Lett.* **2006**, *19*, 1202–1206. [CrossRef]
25. Grace, S.; Graef, J.; Tunc, E. Oscillatory behavior of a third order neutral dynamice equations with distributed delays. *Electron. J. Qual. Theory Differ. Equ.* **2016**, *14*, 1–14.
26. Graef, J.; Grace, S.; Tunc, E. Oscillation criteria for even-order differential equations with unbounded neutral coecients and distributed deviating arguments. *Funct. Differ. Equ.* **2018**, *45*, 143–153.

27. Grace, S.; Graef, J.; Tunc, E. Oscillatory behavior of second order damped neutral differential equations with distributed deviating arguments. *Miskolc Math. Notes* **2017**, *18*, 759–769. [CrossRef]
28. Ozdemir, O.; Tunc, E. Asymptotic behavior and oscillation of solutions of third order neutral dynamic equations with distributed deviating arguments. *Bull. Math. Anal. Appl.* **2018**, *10*, 31–52.

© 2020 by the authors. Licensee MDPI, Basel, Switzerland. This article is an open access article distributed under the terms and conditions of the Creative Commons Attribution (CC BY) license (http://creativecommons.org/licenses/by/4.0/).

Article

Asymptotic Properties of Neutral Differential Equations with Variable Coefficients

Omar Bazighifan [1,2,†], Rami Ahmad El-Nabulsi [3,*,†] and Osama Moaaz [4,†]

1 Department of Mathematics, Faculty of Science, Hadhramout University, Hadhramout 50512, Yemen; o.bazighifan@gmail.com
2 Department of Mathematics, Faculty of of Education, Seiyun University, Hadhramout 50512, Yemen
3 Mathematics and Physics Divisions, Athens Institute for Education and Research, 10671 Athens, Greece
4 Department of Mathematics, Faculty of Science, Mansoura University, Mansoura 35516, Egypt; o_moaaz@mans.edu.eg
* Correspondence: el-nabulsi@atiner.gr
† These authors contributed equally to this work.

Received: 24 April 2020; Accepted: 10 August 2020; Published: 12 August 2020

Abstract: The aim of this work is to study oscillatory behavior of solutions for even-order neutral nonlinear differential equations. By using the Riccati substitution, a new oscillation conditions is obtained which insures that all solutions to the studied equation are oscillatory. The obtained results complement the well-known oscillation results present in the literature. Some example are illustrated to show the applicability of the obtained results.

Keywords: even-order differential equations; neutral delay; oscillation

1. Introduction

Neutral differential equations appear in models concerning biological, physical and chemical phenomena, optimization, mathematics of networks, dynamical systems and their application in concerning materials and energy as well as problems of deformation of structures, elasticity or soil settlement, see [1].

Recently, there has been steady enthusiasm for acquiring adequate conditions for oscillatory and nonoscillatory behavior of differential equations of different orders; see [2–13]. Particular emphasize has been given to the study of oscillation and oscillatory behavior of these equations which have been under investigation by using different methods an various techniques; we refer to the papers [14–26]. In this paper we study the oscillatory behavior of the even-order nonlinear differential equation

$$\left(r(\varsigma) \left(z^{(n-1)}(\varsigma) \right)^{\alpha} \right)' + q(\varsigma) x^{\alpha}(\delta(\varsigma)) = 0, \tag{1}$$

where $\varsigma \geq \varsigma_0$, n is an even natural number and $z(\varsigma) := x(\varsigma) + p(\varsigma) x(\tau(\varsigma))$. Throughout this paper, we suppose that: $r \in C[\varsigma_0, \infty)$, $r(\varsigma) > 0$, $r'(\varsigma) \geq 0$, $p, q \in C([\varsigma_0, \infty))$, $q(\varsigma) > 0$, $0 \leq p(\varsigma) < p_0 < \infty$, q is not identically zero for large ς, $\tau \in C^1[\varsigma_0, \infty)$, $\delta \in C[\varsigma_0, \infty)$, $\tau'(\varsigma) > 0$, $\tau(\varsigma) \leq \varsigma$, $\lim_{\varsigma \to \infty} \tau(\varsigma) = \lim_{\varsigma \to \infty} \delta(\varsigma) = \infty$, α is a quotient of odd positive integers and

$$\int_{\varsigma_0}^{\infty} r^{-1/\alpha}(s) \, ds = \infty. \tag{2}$$

Definition 1. Let x be a real function defined for all ς in a real interval $I := [\varsigma_x, \infty)$, $\varsigma_x \geq \varsigma_0$, and having an $(n-1)^{th}$ derivative for all $\varsigma \in I$. The function f is called a solution of the differential Equation (1) on I if it fulfills the following two requirements:

$$\left(r(\varsigma) \left((x(t) + p(t) x(\tau(t)))^{(n-1)} (\varsigma) \right)^\alpha \right) \in C^1([\varsigma_x, \infty))$$

and

$$x(\varsigma) \text{ satisfies (1) on } [\varsigma_x, \infty).$$

Definition 2. A solution of (1) is called oscillatory if it has arbitrarily large zeros on $[\varsigma_x, \infty)$, and otherwise is called to be nonoscillatory.

Definition 3. The Equation (1) is said to be oscillatory if all its solutions are oscillatory.

We collect some relevant facts and auxiliary results from the existing literature.

Bazighifan [2] using the Riccati transformation together with comparison method with second order equations, focuses on the oscillation of equations of the form

$$\left(r(\varsigma) \left(z^{(n-1)}(\varsigma) \right)^\alpha \right)' + q(\varsigma) f(x(\delta(\varsigma))) = 0, \tag{3}$$

where n is even.

Moaaz et al. [27] gives us some results providing informations on the asymptotic behavior of (1). This time, the authors used comparison method with first-order equations.

In [28] (Theorem 2), the authors considered Equation (1) and proved that (1) is oscillatory if

$$\int_{\varsigma_0}^\infty \left(\Psi(s) - \frac{2^\alpha}{(\alpha+1)^{\alpha+1}} \frac{r(s) (\rho'(s))^{\alpha+1}}{\mu^\alpha s^{2\alpha} \rho^\alpha(s)} \right) ds = \infty,$$

for some $\mu \in (0, 1)$ and

$$\int_{\varsigma_0}^\infty \left(\vartheta(s) \left(\int_s^\infty (Q^*(v))^{\frac{1}{\alpha}} r^{\frac{-1}{\alpha}}(v) dv \right) - \frac{\vartheta_+'^2(s)}{4\vartheta(s)} \right) ds = \infty,$$

where $\Psi(\varsigma) := \vartheta \rho(\varsigma) Q(\varsigma) (1 - p(g(\varsigma, a)))^\alpha (g(\varsigma, a)/\varsigma)^{3\alpha}$.

Xing et al. [29] proved that (1) is oscillatory if

$$\left(\delta^{-1}(\varsigma) \right)' \geq \delta_0 > 0, \ \tau'(\varsigma) \geq \tau_0 > 0, \ \tau^{-1}(\delta(\varsigma)) < \varsigma$$

and

$$\liminf_{\varsigma \to \infty} \int_{\tau^{-1}(\delta(\varsigma))}^\varsigma \frac{\hat{q}(s)}{r(s)} \left(s^{n-1} \right)^\alpha ds > \left(\frac{1}{\delta_0} + \frac{p_0^\alpha}{\delta_0 \tau_0} \right) \frac{((n-1)!)^\alpha}{e},$$

where $\hat{q}(\varsigma) := \min \{ q(\delta^{-1}(\varsigma)), q(\delta^{-1}(\tau(\varsigma))) \}$.

In this article, we establish some oscillation criteria for the Equation (1) which complements some of the results obtained in the literature. Some examples are presented to illustrate our main results.

To prove our main results we need the following lemmas:

Lemma 1 ([28]). Let $\alpha \geq 1$ be a ratio of two odd numbers. Then

$$Dw - Cw^{(\alpha+1)/\alpha} \leq \frac{\alpha^\alpha}{(\alpha+1)^{\alpha+1}} \frac{D^{\alpha+1}}{C^\alpha}, \ C > 0.$$

Lemma 2 ([30])**.** Let $h(\varsigma) \in C^n([\varsigma_0, \infty), (0, \infty))$. If $h^{(n-1)}(\varsigma) h^{(n)}(\varsigma) \leq 0$ for all $\varsigma \geq \varsigma_x$, then for every $\theta \in (0, 1)$, there exists a constant $M > 0$ such that

$$h'(\theta\varsigma) \geq M\varsigma^{n-2}h^{(n-1)}(\varsigma),$$

for all sufficient large ς.

Lemma 3 ([31] Lemma 2.2.3)**.** Let $x \in C^n([\varsigma_0, \infty), (0, \infty))$. Assume that $x^{(n)}(\varsigma)$ is of fixed sign and not identically zero on $[\varsigma_0, \infty)$ and that there exists a $\varsigma_1 \geq \varsigma_0$ such that $x^{(n-1)}(\varsigma) x^{(n)}(\varsigma) \leq 0$ for all $\varsigma \geq \varsigma_1$. If $\lim_{\varsigma \to \infty} x(\varsigma) \neq 0$, then for every $\mu \in (0, 1)$ there exists $\varsigma_\mu \geq \varsigma_1$ such that

$$x(\varsigma) \geq \frac{\mu}{(n-1)!} \varsigma^{n-1} \left| x^{(n-1)}(\varsigma) \right| \text{ for } \varsigma \geq \varsigma_\mu.$$

Lemma 4 ([32])**.** Let $h \in C^n([\varsigma_0, \infty), (0, \infty))$. If $h^{(n)}(\varsigma)$ is eventually of one sign for all large ς, then there exists a $\varsigma_x > \varsigma_1$ for some $\varsigma_1 > \varsigma_0$ and an integer m, $0 \leq m \leq n$ with $n + m$ even for $h^{(n)}(\varsigma) \geq 0$ or $n + m$ odd for $h^{(n)}(\varsigma) \leq 0$ such that $m > 0$ implies that $h^{(k)}(\varsigma) > 0$ for $\varsigma > \varsigma_x$, $k = 0, 1, \ldots, m-1$ and $m \leq n - 1$ implies that $(-1)^{m+k} h^{(k)}(\varsigma) > 0$ for $\varsigma > \varsigma_x, k = m, m+1, \ldots, n-1$.

2. One Condition Theorem

Notation 1. Here, we define the next notation:

$$\Omega(s) = \frac{\vartheta(s)}{\delta_0(\alpha+1)^{\alpha+1}(\lambda M)^\alpha} \left(\varphi(s) + \frac{\vartheta'(s)}{\vartheta(s)} \right)^{\alpha+1},$$

$$\Theta(s) = \frac{\vartheta(s)((n-2)!)^\alpha}{\mu^\alpha \delta_0(\alpha+1)^{\alpha+1}} \left(\varphi(s) + \frac{\vartheta'(s)}{\vartheta(s)} \right)^{\alpha+1}$$

and

$$Q(s) = \min\left\{ q\left(\delta^{-1}(s)\right), q\left(\delta^{-1}(\tau(s))\right) \right\}.$$

Following [33], we say that a function $\Phi = \Phi(\varsigma, s, l)$ belongs to the function class Y if $\Phi \in (E, \mathbb{R})$ where $E = \{(\varsigma, s, l) : \varsigma_0 \leq 1 \leq s \leq \varsigma\}$ which satisfies $\Phi(\varsigma, \varsigma, l) = 0$, $\Phi(\varsigma, l, l) = 0$ and $\Phi(\varsigma, s, l) > 0$, for $l < s < \varsigma$ and has the partial derivative $\partial\Phi/\partial s$ on E such that $\partial\Phi/\partial s$ is locally integrable with respect to s in E.

Definition 4. Define the operator $B[\cdot; l, \varsigma]$ by

$$B[h; l, \varsigma] = \int_l^\varsigma \Phi(\varsigma, s, l) h(s) \, ds,$$

for $\varsigma_0 \leq 1 \leq s \leq \varsigma$ and $h \in C([\varsigma_0, \infty), \mathbb{R})$. The function $\varphi = \varphi(\varsigma, s, l)$ is defined by

$$\frac{\partial \Phi(\varsigma, s, l)}{\partial s} = \varphi(\varsigma, s, l) \Phi(\varsigma, s, l).$$

Remark 1. It is easy to verify that $B[\cdot; l, \varsigma]$ is a linear operator and that it satisfies

$$B[h'; l, \varsigma] = -B[h\varphi; l, \varsigma], \text{ for } h \in C^1([\varsigma_0, \infty), \mathbb{R}). \tag{4}$$

Lemma 5. Assume that x is an eventually positive solution of (1) and

$$\left(\delta^{-1}(\varsigma)\right)' \geq \delta_0 > 0, \quad (\tau(\varsigma))' \geq \tau_0 > 0. \tag{5}$$

Then

$$\frac{1}{\delta_0}\left(r\left(\delta^{-1}(\varsigma)\right)\left(z^{(n-1)}\left(\delta^{-1}(\varsigma)\right)\right)^{\alpha}\right)'$$
$$+\frac{p_0^{\alpha}}{\delta_0\tau_0}\left(r\left(\delta^{-1}(\tau(\varsigma))\right)\left(z^{(n-1)}\left(\delta^{-1}(\tau(\varsigma))\right)\right)^{\alpha}\right)' + Q(\varsigma)z^{\alpha}(\varsigma) \leq 0. \quad (6)$$

Proof. Let x be an eventually positive solution of (1) on $[\varsigma_0, \infty)$. From (1), we see that

$$0 = \frac{1}{(\delta^{-1}(\varsigma))'}\left(r\left(\delta^{-1}(\varsigma)\right)\left(z^{(n-1)}\left(\delta^{-1}(\varsigma)\right)\right)^{\alpha}\right)' + q\left(\delta^{-1}(\varsigma)\right)x^{\alpha}(\varsigma). \quad (7)$$

Thus, for all sufficiently large ς, we have

$$0 = \frac{1}{(\delta^{-1}(\varsigma))'}\left(r\left(\delta^{-1}(\varsigma)\right)\left(z^{(n-1)}\left(\delta^{-1}(\varsigma)\right)\right)^{\alpha}\right)'$$
$$+ \frac{p_0^{\alpha}}{(\delta^{-1}(\tau(\varsigma)))'}\left(r\left(\delta^{-1}(\tau(\varsigma))\right)\left(z^{(n-1)}\left(\delta^{-1}(\tau(\varsigma))\right)\right)^{\alpha}\right)'$$
$$+ q\left(\delta^{-1}(\varsigma)\right)x^{\alpha}(\varsigma) + p_0^{\alpha}q\left(\delta^{-1}(\tau(\varsigma))\right)x^{\alpha}(\tau(\varsigma)). \quad (8)$$

From (8) and the definition of z, we get

$$q\left(\delta^{-1}(\varsigma)\right)x^{\alpha}(\varsigma) + p_0^{\alpha}q\left(\delta^{-1}(\tau(\varsigma))\right)x^{\alpha}(\tau(\varsigma)) \geq Q(\varsigma)(x(\varsigma) + p_0x(\tau(\varsigma)))^{\alpha}$$
$$\geq Q(\varsigma)z^{\alpha}(\varsigma). \quad (9)$$

Thus, by using (8) and (9), we obtain

$$0 \geq \frac{1}{(\delta^{-1}(\varsigma))'}\left(r\left(\delta^{-1}(\varsigma)\right)\left(z^{(n-1)}\left(\delta^{-1}(\varsigma)\right)\right)^{\alpha}\right)'$$
$$+ \frac{p_0^{\alpha}}{(\delta^{-1}(\tau(\varsigma)))'}\left(r\left(\delta^{-1}(\tau(\varsigma))\right)\left(z^{(n-1)}\left(\delta^{-1}(\tau(\varsigma))\right)\right)^{\alpha}\right)' + Q(\varsigma)z^{\alpha}(\varsigma). \quad (10)$$

From (5), we get

$$0 \geq \frac{1}{\delta_0}\left(r\left(\delta^{-1}(\varsigma)\right)\left(z^{(n-1)}\left(\delta^{-1}(\varsigma)\right)\right)^{\alpha}\right)'$$
$$+ \frac{p_0^{\alpha}}{\delta_0\tau_0}\left(r\left(\delta^{-1}(\tau(\varsigma))\right)\left(z^{(n-1)}\left(\delta^{-1}(\tau(\varsigma))\right)\right)^{\alpha}\right)' + Q(\varsigma)z^{\alpha}(\varsigma).$$

This completes the proof. □

Theorem 1. *Let (2) hold. Assume that there exist positive functions $\vartheta \in C^1([\varsigma_0, \infty), \mathbb{R})$ such that for all $M > 0$*

$$\limsup_{\varsigma \to \infty} B\left[\vartheta(s)Q(s) - \Omega(s)\left(\frac{r(\delta^{-1}(s))}{\left((\delta^{-1}(s))^{n-2}\right)^{\alpha}} + \frac{p_0^{\alpha}r(\delta^{-1}(\tau(s)))}{\tau_0\left((\delta^{-1}(\tau(s)))^{n-2}\right)^{\alpha}}\right); l, \varsigma\right] > 0, \quad (11)$$

for some $\lambda \in (0,1)$, then (1) is oscillatory.

Proof. Suppose that (1) has a nonoscillatory solution in $[\varsigma_0, \infty)$. Without loss of generality, we let x be an eventually positive solution of (1). Then, there exists a $\varsigma_1 \geq \varsigma_0$ such that $x(\varsigma) > 0$, $x(\tau(\varsigma)) > 0$ and $x(\delta(\varsigma)) > 0$ for $\varsigma \geq \varsigma_1$. Thus, we have

$$z(\varsigma) > 0,\ z'(\varsigma) > 0,\ z^{(n-1)}(\varsigma) > 0,\ z^{(n)}(\varsigma) < 0. \tag{12}$$

By Lemma 2, we get

$$z'(\lambda\varsigma) \geq M\varsigma^{n-2} z^{(n-1)}(\varsigma), \tag{13}$$

where M is positive constant. Now, we define a function ψ by

$$\psi(\varsigma) = \vartheta(\varsigma) \frac{r\left(\delta^{-1}(\varsigma)\right)\left(z^{(n-1)}\left(\delta^{-1}(\varsigma)\right)\right)^{\alpha}}{z^{\alpha}(\lambda\varsigma)}. \tag{14}$$

Then we obtain $\psi(\varsigma) > 0$ for $\varsigma \geq \varsigma_1$, and

$$\begin{aligned}\psi'(\varsigma) &= \vartheta'(\varsigma) \frac{r\left(\delta^{-1}(\varsigma)\right)\left(z^{(n-1)}\left(\delta^{-1}(\varsigma)\right)\right)^{\alpha}}{z^{\alpha}(\lambda\varsigma)} + \vartheta(\varsigma) \frac{\left(r\left(\delta^{-1}(\varsigma)\right)\left(z^{(n-1)}\left(\delta^{-1}(\varsigma)\right)\right)^{\alpha}\right)'}{z^{\alpha}(\lambda\varsigma)} \\ &\quad - \alpha\lambda\vartheta(\varsigma) \frac{r\left(\delta^{-1}(\varsigma)\right)\left(z^{(n-1)}\left(\delta^{-1}(\varsigma)\right)\right)^{\alpha} z'(\lambda\varsigma)}{z^{\alpha+1}(\lambda\varsigma)}. \end{aligned} \tag{15}$$

Combining (13) and (14) in (15), we obtain

$$\begin{aligned}\psi'(\varsigma) &\leq \frac{\vartheta'(\varsigma)}{\vartheta(\varsigma)} \psi(\varsigma) + \vartheta(\varsigma) \frac{\left(r\left(\delta^{-1}(\varsigma)\right)\left(z^{(n-1)}\left(\delta^{-1}(\varsigma)\right)\right)^{\alpha}\right)'}{z^{\alpha}(\lambda\varsigma)} \\ &\quad - \alpha\lambda M \left(\delta^{-1}(\varsigma)\right)^{n-2} \frac{(\psi(\varsigma))^{\alpha+1/\alpha}}{\left(\vartheta(\varsigma) r\left(\delta^{-1}(\varsigma)\right)\right)^{1/\alpha}}. \end{aligned} \tag{16}$$

Similarly, define

$$\tilde{\psi}(\varsigma) = \vartheta(\varsigma) \frac{r\left(\delta^{-1}(\tau(\varsigma))\right)\left(z^{(n-1)}\left(\delta^{-1}(\tau(\varsigma))\right)\right)^{\alpha}}{z^{\alpha}(\lambda\varsigma)}. \tag{17}$$

Then we obtain $\tilde{\psi}(\varsigma) > 0$ for $\varsigma \geq \varsigma_1$, and

$$\begin{aligned}\tilde{\psi}'(\varsigma) &\leq \frac{\vartheta'(\varsigma)}{\vartheta(\varsigma)} \tilde{\psi}(\varsigma) + \vartheta(\varsigma) \frac{\left(r\left(\delta^{-1}(\tau(\varsigma))\right)\left(z^{(n-1)}\left(\delta^{-1}(\tau(\varsigma))\right)\right)^{\alpha}\right)'}{z^{\alpha}(\lambda\varsigma)} \\ &\quad - \alpha\lambda M \left(\delta^{-1}(\tau(\varsigma))\right)^{n-2} \frac{(\tilde{\psi}(\varsigma))^{\alpha+1/\alpha}}{\left(\vartheta(\varsigma) r\left(\delta^{-1}(\tau(\varsigma))\right)\right)^{1/\alpha}}. \end{aligned} \tag{18}$$

Therefore, from (16) and (18), we obtain

$$\begin{aligned}\frac{1}{\delta_0}\psi'(\varsigma) + \frac{p_0^{\alpha}}{\delta_0 \tau_0}\tilde{\psi}'(\varsigma) &\leq \frac{\vartheta(\varsigma)}{\delta_0} \frac{\left(r\left(\delta^{-1}(\varsigma)\right)\left(z^{(n-1)}\left(\delta^{-1}(\varsigma)\right)\right)^{\alpha}\right)'}{z^{\alpha}(\lambda\varsigma)} \\ &\quad + \frac{p_0^{\alpha}}{\delta_0 \tau_0} \vartheta(\varsigma) \frac{\left(r\left(\delta^{-1}(\tau(\varsigma))\right)\left(z^{(n-1)}\left(\delta^{-1}(\tau(\varsigma))\right)\right)^{\alpha}\right)'}{z^{\alpha}(\lambda\varsigma)} \\ &\quad + \frac{\vartheta'(\varsigma)}{\delta_0 \vartheta(\varsigma)} \psi(\varsigma) + \frac{p_0^{\alpha}}{\delta_0 \tau_0} \frac{\vartheta'(\varsigma)}{\vartheta(\varsigma)} \tilde{\psi}(\varsigma) \\ &\quad - \frac{1}{\delta_0} \alpha\lambda M \left(\delta^{-1}(\varsigma)\right)^{n-2} \frac{(\psi(\varsigma))^{\alpha+1/\alpha}}{\left(\vartheta(\varsigma) r\left(\delta^{-1}(\varsigma)\right)\right)^{1/\alpha}} \\ &\quad - \alpha\lambda M \frac{p_0^{\alpha}}{\delta_0 \tau_0} \left(\delta^{-1}(\tau(\varsigma))\right)^{n-2} \frac{(\tilde{\psi}(\varsigma))^{\alpha+1/\alpha}}{\left(\vartheta(\varsigma) r\left(\delta^{-1}(\tau(\varsigma))\right)\right)^{1/\alpha}}. \end{aligned} \tag{19}$$

From (16), we obtain

$$\frac{1}{\delta_0}\psi'(\varsigma) + \frac{p_0^\alpha}{\delta_0\tau_0}\tilde{\psi}'(\varsigma) \leq -\vartheta(\varsigma)Q(\varsigma) + \frac{\vartheta'(\varsigma)}{\delta_0\vartheta(\varsigma)}\psi(\varsigma) + \frac{p_0^\alpha}{\delta_0\tau_0}\frac{\vartheta'(\varsigma)}{\vartheta(\varsigma)}\tilde{\psi}(\varsigma)$$
$$-\frac{1}{\delta_0}\alpha\lambda M\left(\delta^{-1}(\varsigma)\right)^{n-2}\frac{(\psi(\varsigma))^{\alpha+1/\alpha}}{(\vartheta(\varsigma)r(\delta^{-1}(\varsigma)))^{1/\alpha}}$$
$$-\alpha\lambda M\frac{p_0^\alpha}{\delta_0\tau_0}\left(\delta^{-1}(\tau(\varsigma))\right)^{n-2}\frac{(\tilde{\psi}(\varsigma))^{\alpha+1/\alpha}}{(\vartheta(\varsigma)r(\delta^{-1}(\tau(\varsigma))))^{1/\alpha}}. \quad (20)$$

Applying $B[\cdot;l,\varsigma]$ to (20), we obtain

$$B\left[\frac{1}{\delta_0}\psi'(\varsigma) + \frac{p_0^\alpha}{\delta_0\tau_0}\tilde{\psi}'(\varsigma);l,\varsigma\right] \leq B[-\vartheta(s)Q(s) + \frac{\vartheta'(s)}{\delta_0\vartheta(s)}\psi(s) + \frac{p_0^\alpha}{\delta_0\tau_0}\frac{\vartheta'(s)}{\vartheta(s)}\tilde{\psi}(s)$$
$$-\frac{1}{\delta_0}\alpha\lambda M\left(\delta^{-1}(s)\right)^{n-2}\frac{(\psi(s))^{\alpha+1/\alpha}}{(\vartheta(s)r(\delta^{-1}(s)))^{1/\alpha}} \quad (21)$$
$$-\alpha\lambda M\frac{p_0^\alpha}{\delta_0\tau_0}\left(\delta^{-1}(\tau(s))\right)^{n-2}\frac{(\tilde{\psi}(s))^{\alpha+1/\alpha}}{(\vartheta(s)r(\delta^{-1}(\tau(s))))^{1/\alpha}};l,\varsigma].$$

By (4) and the inequality above, we find

$$\varsigma[\vartheta(s)Q(s);l,\varsigma] \leq \varsigma[\frac{1}{\delta_0}\left(\varphi(s) + \frac{\vartheta'(s)}{\vartheta(s)}\right)\psi(s) + \frac{p_0^\alpha}{\delta_0\tau_0}\left(\varphi(s) + \frac{\vartheta'(s)}{\vartheta(s)}\right)\tilde{\psi}(s)$$
$$-\frac{1}{\delta_0}\alpha\lambda M\left(\delta^{-1}(s)\right)^{n-2}\frac{(\psi(s))^{\alpha+1/\alpha}}{(\vartheta(s)r(\delta^{-1}(s)))^{1/\alpha}}$$
$$-\alpha\lambda M\frac{p_0^\alpha}{\delta_0\tau_0}\left(\delta^{-1}(\tau(s))\right)^{n-2}\frac{(\tilde{\psi}(s))^{\alpha+1/\alpha}}{(\vartheta(s)r(\delta^{-1}(\tau(s))))^{1/\alpha}};l,\varsigma]. \quad (22)$$

Using Lemma 1, we set

$$D = \frac{1}{\delta_0}\left(\varphi(s) + \frac{\vartheta'(s)}{\vartheta(s)}\right), \quad C = \frac{\frac{1}{\delta_0}\alpha\lambda M\left(\delta^{-1}(s)\right)^{n-2}}{(\vartheta(s)r(\delta^{-1}(s)))^{1/\alpha}} \quad \text{and} \quad w = \psi,$$

we have

$$\frac{1}{\delta_0}\left(\varphi(s) + \frac{\vartheta'(s)}{\vartheta(s)}\right)\psi(s) - \frac{1}{\delta_0}\alpha\lambda M\left(\delta^{-1}(s)\right)^{n-2}\frac{(\psi(s))^{\alpha+1/\alpha}}{(\vartheta(\varsigma)r(\delta^{-1}(s)))^{1/\alpha}}$$
$$< \frac{1}{(\alpha+1)^{\alpha+1}\delta_0}\left(\varphi(\varsigma) + \frac{\vartheta'(s)}{\vartheta(s)}\right)^{\alpha+1}\frac{\vartheta(\varsigma)r(\delta^{-1}(\tau(s)))}{\left(\lambda M\left(\delta^{-1}(s)\right)^{n-2}\right)^\alpha}. \quad (23)$$

Hence, from (22) and (23), we have

$$B[\vartheta(s)Q(s);l,\varsigma] \leq B[\left(\varphi(s) + \frac{\vartheta'(s)}{\vartheta(s)}\right)^{\alpha+1}\frac{\vartheta(s)r(\delta^{-1}(s))}{(\alpha+1)^{\alpha+1}\delta_0\left(\lambda M\left(\delta^{-1}(s)\right)^{n-2}\right)^\alpha}$$
$$+ \left(\varphi(s) + \frac{\vartheta'(s)}{\vartheta(s)}\right)^{\alpha+1}\frac{p_0^\alpha\vartheta(s)r(\delta^{-1}(\tau(s)))}{(\alpha+1)^{\alpha+1}\delta_0\tau_0\left(\lambda M\left(\delta^{-1}(\tau(s))\right)^{n-2}\right)^\alpha};l,\varsigma].$$

Easily, we find that

$$B\left[\vartheta(s)Q(s);l,\varsigma\right] \leq B\left[\frac{\vartheta(s)}{\delta_0(\alpha+1)^{\alpha+1}(\lambda M)^\alpha}\left(\varphi(s)+\frac{\vartheta'(s)}{\vartheta(s)}\right)^{\alpha+1}\right.$$
$$\left.\times\left(\frac{r\left(\delta^{-1}(s)\right)}{\left(\left(\delta^{-1}(s)\right)^{n-2}\right)^\alpha}+\frac{p_0^\alpha r\left(\delta^{-1}(\tau(s))\right)}{\tau_0\left(\left(\delta^{-1}(\tau(s))\right)^{n-2}\right)^\alpha}\right);l,\varsigma\right].$$

That is,

$$B\left[\vartheta(s)Q(s)-\Omega(\varsigma)\left(\frac{r\left(\delta^{-1}(s)\right)}{\left(\left(\delta^{-1}(s)\right)^{n-2}\right)^\alpha}+\frac{p_0^\alpha r\left(\delta^{-1}(\tau(s))\right)}{\tau_0\left(\left(\delta^{-1}(\tau(s))\right)^{n-2}\right)^\alpha}\right);l,\varsigma\right] \leq 0.$$

Taking the super limit in the inequality above, we obtain

$$\limsup_{\varsigma\to\infty} B\left[\vartheta(s)Q(s)-\Omega(s)\left(\frac{r\left(\delta^{-1}(s)\right)}{\left(\left(\delta^{-1}(s)\right)^{n-2}\right)^\alpha}+\frac{p_0^\alpha r\left(\delta^{-1}(\tau(s))\right)}{\tau_0\left(\left(\delta^{-1}(\tau(s))\right)^{n-2}\right)^\alpha}\right);l,\varsigma\right] \leq 0, \quad (24)$$

which is a contradiction. The proof is complete. □

3. Tow Conditions Theorem

Lemma 6 ([22]). *(Lemma 1.2) Assume that x is an eventually positive solution of (1). Then, there exists two possible cases:*

(I_1) $z(\varsigma)>0, z'(\varsigma)>0, z''(\varsigma)>0, z^{(n-1)}(\varsigma)>0, z^{(n)}(\varsigma)<0$,

(I_2) $z(\varsigma)>0, z^{(j)}(\varsigma)>0, z^{(j+1)}(\varsigma)<0$ *for all odd integer*
$j\in\{1,2,\ldots,n-3\}, z^{(n-1)}(\varsigma)>0, z^{(n)}(\varsigma)<0$,

for $\varsigma\geq\varsigma_1$, where $\varsigma_1\geq\varsigma_0$ is sufficiently large.

Lemma 7 ([22]). *(Lemma 1.2) Assume that x is an eventually positive solution of (1) and*

$$\int_{\varsigma_0}^\infty\left(\Psi(s)-\frac{2^\alpha}{(\alpha+1)^{\alpha+1}}\frac{r(s)\left(\rho'(s)\right)^{\alpha+1}}{\mu^\alpha s^{2\alpha}\rho^\alpha(s)}\right)ds=\infty, \quad (25)$$

where

$$\Psi(\varsigma)=\vartheta\rho(\varsigma)q(\varsigma)\left(1-p(\delta(\varsigma))\right)^\alpha\left(\delta(\varsigma)\backslash\varsigma\right)^{3\alpha},$$

where $\rho\in C^1\left([\varsigma_0,\infty),(0,\infty)\right)$, then it will be z does not satisfy case (I_1).

Lemma 8. *Let (2) holds and assume that x is an eventually positive solution of (1). If there exists positive functions $\vartheta\in C^1\left([\varsigma_0,\infty),\mathbb{R}\right)$ such that for all $M>0$*

$$\limsup_{\varsigma\to\infty} B\left[\vartheta(s)Q(s)-\Theta(s)\left(\frac{r\left(\delta^{-1}(s)\right)}{\left(\left(\delta^{-1}(s)\right)^{n-2}\right)^\alpha}+\frac{p_0^\alpha r\left(\delta^{-1}(\tau(s))\right)}{\tau_0\left(\left(\delta^{-1}(\tau(s))\right)^{n-2}\right)^\alpha}\right);l,\varsigma\right]>0, \quad (26)$$

for some $\mu\in(0,1)$, then z not satisfies case (I_2).

Proof. Assume to the contrary that (1) has a nonoscillatory solution in $[\varsigma_0,\infty)$. Without loss of generality, we let x be an eventually positive solution of (1). From Lemma 3, we obtain

$$z'(\varsigma)\geq\frac{\mu}{(n-2)!}\varsigma^{n-2}z^{(n-1)}(\varsigma). \quad (27)$$

Now, we define a function ω by

$$\omega(\varsigma) = \vartheta(\varsigma) \frac{r(\delta^{-1}(\varsigma)) \left(z^{(n-1)}(\delta^{-1}(\varsigma))\right)^{\alpha}}{z^{\alpha}(\varsigma)}. \tag{28}$$

Then we see that $\omega(\varsigma) > 0$ for $\varsigma \geq \varsigma_1$, and

$$\omega'(\varsigma) \leq \frac{\vartheta'(\varsigma)}{\vartheta(\varsigma)} \omega(\varsigma) + \vartheta(\varsigma) \frac{\left(r(\delta^{-1}(\varsigma)) \left(z^{(n-1)}(\delta^{-1}(\varsigma))\right)^{\alpha}\right)'}{z^{\alpha}(\lambda \varsigma)}$$
$$- \alpha \frac{\mu}{(n-2)!} \left(\delta^{-1}(\varsigma)\right)^{n-2} \frac{(\omega(\varsigma))^{\alpha+1/\alpha}}{(\vartheta(\varsigma) r(\delta^{-1}(\varsigma)))^{1/\alpha}}. \tag{29}$$

Similarly, define

$$\tilde{\omega}(\varsigma) = \vartheta(\varsigma) \frac{r(\delta^{-1}(\tau(\varsigma))) \left(z^{(n-1)}(\delta^{-1}(\tau(\varsigma)))\right)^{\alpha}}{z^{\alpha}(\varsigma)}. \tag{30}$$

Then we see that $\tilde{\omega}(\varsigma) > 0$ for $\varsigma \geq \varsigma_1$, and

$$\tilde{\omega}'(\varsigma) \leq \frac{\vartheta'(\varsigma)}{\vartheta(\varsigma)} \tilde{\omega}(\varsigma) + \vartheta(\varsigma) \frac{\left(r(\delta^{-1}(\tau(\varsigma))) \left(z^{(n-1)}(\delta^{-1}(\tau(\varsigma)))\right)^{\alpha}\right)'}{z^{\alpha}(\varsigma)}$$
$$- \alpha \frac{\mu}{(n-2)!} \left(\delta^{-1}(\tau(\varsigma))\right)^{n-2} \frac{(\tilde{\omega}(\varsigma))^{\alpha+1/\alpha}}{(\vartheta(\varsigma) r(\delta^{-1}(\tau(\varsigma))))^{1/\alpha}}.$$

Thus, we get

$$\limsup_{\varsigma \to \infty} B \left[\vartheta(s) Q(s) - \Theta(s) \left(\frac{r(\delta^{-1}(s))}{\left((\delta^{-1}(s))^{n-2}\right)^{\alpha}} + \frac{p_0^{\alpha} r(\delta^{-1}(\tau(s)))}{\tau_0 \left((\delta^{-1}(\tau(s)))^{n-2}\right)^{\alpha}} \right); l, \varsigma \right] \leq 0,$$

which is a contradiction. The proof is complete. □

Theorem 2. *Assume that (25) and (26) hold for some $\mu \in (0,1)$. Then every solution of (1) is oscillatory.*

Example 1. *Consider the equation*

$$(x(\varsigma) + 2x(\varsigma - 5\pi))'' + q_0 x(\varsigma - \pi) = 0. \tag{31}$$

We note that $r(\varsigma) = 1$, $p(\varsigma) = 2$, $\tau(\varsigma) = \varsigma - 5\pi$, $\delta(\varsigma) = \varsigma - \pi$, $\delta^{-1}(s) = \varsigma + \pi$ and $q(\varsigma) = Q(\varsigma) = q_0$. Thus, if we choose $\Phi(\varsigma) = (\varsigma - s)(s - l)$, then it is easy to see that

$$\varphi(\varsigma, s, l) = \frac{(\varsigma - s) - (s - l)}{(\varsigma - s)(s - l)}$$

and

$$\Omega(s) = \frac{\vartheta(s)}{\delta_0(\alpha+1)^{\alpha+1}(\lambda M)^{\alpha}} \left(\varphi(s) + \frac{\vartheta'(s)}{\vartheta(s)} \right)^{\alpha+1}$$
$$= \frac{1}{4\lambda M} \left(\frac{(\varsigma - s) - (s - l)}{(\varsigma - s)(s - l)} \right)^2.$$

Thus,

$$\limsup_{\varsigma \to \infty} B\left[\vartheta(s) Q(s) - \Omega(s) \left(\frac{r(\delta^{-1}(s))}{\left((\delta^{-1}(s))^{n-2}\right)^\alpha} + \frac{p_0^\alpha r(\delta^{-1}(\tau(s)))}{\tau_0 \left((\delta^{-1}(\tau(s)))^{n-2}\right)^\alpha} \right); l, \varsigma \right]$$
$$= \limsup_{\varsigma \to \infty} B\left[q_0 - \frac{3}{4\lambda M} \left(\frac{(\varsigma-s)-(s-l)}{(\varsigma-s)(s-l)} \right)^2; l, \varsigma \right] > 0.$$

Therefore, by Theorem 1, every solution of Equation (31) is oscillatory.

Example 2. *Consider the equation*

$$(x(\varsigma) + p_0 x(\varsigma - 5\pi))^{(4)} + q_0 x(\varsigma - \pi), \qquad (32)$$

where $q_0 > 0$. Let $r(\varsigma) = 1$, $p(\varsigma) = p_0$, $\tau(\varsigma) = \varsigma - 5\pi$, $\delta(\varsigma) = \varsigma - \pi$, $\delta^{-1}(s) = \varsigma + \pi$ and $q(\varsigma) = Q(\varsigma) = q_0$, then we have

$$\int_{\varsigma_0}^{\infty} r^{-1/\alpha}(s) \, ds = \infty.$$

Next, if we choose $\varphi(\varsigma) = (\varsigma - s)(s - l)$, then we conclude that the conditions (25) and (26) are satisfied. Thus, using Theorem 2, Equation (32) is oscillatory.

4. Conclusions

In this work, by using the generalized Riccati transformations technique, we provided new oscillation criteria for (1). Furthermore, in future work, by using the comparison method, we find some new Hille and Nehari types and Philos type oscillation criteria of (1).

Author Contributions: Writing original draft, formal analysis, writing review and editing, O.B. and O.M.; writing review and editing, funding and supervision, R.A.E.-N. All authors have read and agreed to the published version of the manuscript.

Funding: This research received no external funding.

Acknowledgments: The authors thank the reviewers for for their useful comments, which led to the improvement of the content of the paper.

Conflicts of Interest: The authors declare no conflict of interest.

References

1. Hale, J.K. *Theory of Functional Differential Equations*; Springer: New York, NY, USA, 1977.
2. Bazighifan, O. An Approach for Studying Asymptotic Properties of Solutions of Neutral Differential Equations. *Symmetry* **2020**, *12*, 555. [CrossRef]
3. Agarwal, R.P.; Bohner, M.; Li, T.; Zhang, C. A new approach in the study of oscillatory behavior of even-order neutral delay diferential equations. *Appl. Math. Comput.* **2013**, *225*, 787–794.
4. Agarwal, R.; Grace, S.; O'Regan, D. *Oscillation Theory for Difference and Functional Differential Equations*; Kluwer Academic Publishers: Dordrecht, The Netherlands, 2000.
5. Baculikova, B.; Dzurina, J. Oscillation theorems for second-order nonlinear neutral differential equations. *Comput. Math. Appl.* **2011**, *62*, 4472–4478. [CrossRef]
6. Baculikova, B.; Dzurina, J.; Li, T. Oscillation results for even-order quasi linear neutral functional differential equations. *Electron. J. Differ. Equ.* **2011**, *2011*, 1–9.
7. Baculikova, B.; Dzurina, J. Oscillation theorems for higher order neutral diferential equations. *Appl. Math. Comput.* **2012**, *219*, 3769–3778.

8. Bazighifan, O. Kamenev and Philos-types oscillation criteria for fourth-order neutral differential equations. *Adv. Differ. Equ.* **2020**, *201*, 1–12.
9. Bazighifan, O.; Ramos, H. On the asymptotic and oscillatory behavior of the solutions of a class of higher-order differential equations with middle term. *Appl. Math. Lett.* **2020**, *107*, 106431. [CrossRef]
10. Bazighifan, O.; Chatzarakis, G.E. Oscillatory and asymptotic behavior of advanced differential equations. *Adv. Differ. Equ.* **2020**, *2020*, 414. [CrossRef]
11. Bazighifan, O.; Postolache, M. An improved conditions for oscillation of functional nonlinear differential equations. *Mathematics* **2020**, *8*, 552. [CrossRef]
12. Bazighifan, O. On the oscillation of certain fourth-order differential equations with p-Laplacian like operator. *Appl. Math. Comput.* **2020**, *386*, 125475. [CrossRef]
13. Bazighifan, O. Oscillatory applications of some fourth-order differential equations. *Math. Meth. Appl. Sci.* **2020**, 1–11. [CrossRef]
14. Li, T.; Han, Z.; Zhao, P.; Sun, S. Oscillation of even-order neutral delay differential equations. *Adv. Differ. Equ.* **2010**, *2010*, 1–9. [CrossRef]
15. Kiguradze, I.T.; Chanturiya, T.A. *Asymptotic Properties of Solutions of Nonautonomous Ordinary Differential Equations*; Kluwer Academic Publishers: Dordrecht, The Netherlands, 1993.
16. Moaaz, O.; Dassios, I.; Bazighifan, O.; Muhib, A. Oscillation Theorems for Nonlinear Differential Equations of Fourth-Order. *Mathematics* **2020**, *8*, 520. [CrossRef]
17. Moaaz, O.; Furuichi, S.; Muhib, A. New Comparison Theorems for the Nth Order Neutral Differential Equations with Delay Inequalities. *Mathematics* **2020**, *8*, 454. [CrossRef]
18. Moaaz, O.; Elabbasy, E.M.; Bazighifan, O. On the asymptotic behavior of fourth-order functional differential equations. *Adv. Differ. Equ.* **2017**, *2017*, 261. [CrossRef]
19. Moaaz, O.; Awrejcewicz, J.; Bazighifan, O. A New Approach in the Study of Oscillation Criteria of Even-Order Neutral Differential Equations. *Mathematics* **2020**, *8*, 197. [CrossRef]
20. Moaaz, O.; Dassios, I.; Bazighifan, O. Oscillation Criteria of Higher-order Neutral Differential Equations with Several Deviating Arguments. *Mathematics* **2020**, *8*, 412. [CrossRef]
21. Meng, F.; Xu, R. Oscillation criteria for certain even order quasi-linear neutral differential equations with deviating arguments. *Appl. Math. Comput.* **2007**, *190*, 458–464. [CrossRef]
22. Philos, C.G. A new criterion for the oscillatory and asymptotic behavior of delay differential equations. *Bull. Acad. Pol. Sci. Ser. Sci. Math.* **1981**, *39*, 61–64.
23. Park, C.; Moaaz, O.; Bazighifan, O. Oscillation Results for Higher Order Differential Equations. *Axioms* **2020**, *9*, 14. [CrossRef]
24. Zafer, A. Oscillation criteria for even order neutral differential equations. *Appl. Math. Lett.* **1998**, *11*, 21–25. [CrossRef]
25. Elabbasy, E.M.; Cesarano, C.; Bazighifan, O.; Moaaz, O. Asymptotic and oscillatory behavior of solutions of a class of higher order differential equation. *Symmetry* **2019**, *11*, 1434. [CrossRef]
26. El-Nabulsi, R.A.; Moaaz, O.; Bazighifan, O. New Results for Oscillatory Behavior of Fourth-Order Differential Equations. *Symmetry* **2020**, *12*, 136. [CrossRef]
27. Moaaz, O.; Kumam, P.; Bazighifan, O. On the Oscillatory Behavior of a Class of Fourth-Order Nonlinear Differential Equation. *Symmetry* **2020**, *12*, 524. [CrossRef]
28. Chatzarakis, G.E.; Elabbasy, E.M.; Bazighifan, O. An oscillation criterion in 4th-order neutral differential equations with a continuously distributed delay. *Adv. Differ. Equ.* **2019**, *336*, 1–9.
29. Xing, G.; Li, T.; Zhang, C. Oscillation of higher-order quasi linear neutral differential equations. *Adv. Differ. Equ.* **2011**, *2011*, 1–10. [CrossRef]
30. Moaaz, O.; Elabbasy, E.M.; Muhib, A. Oscillation criteria for even-order neutral differential equations with distributed deviating arguments. *Adv. Differ. Equ.* **2019**, *297*, 1–10. [CrossRef]
31. Philos, C.G. On the existence of non-oscillatory solutions tending to zero at ∞ for differential equations with positive delays. *Arch. Math.* **1981**, *36*, 168–178. [CrossRef]

32. Zhang, Q.; Yan, J. Oscillation behavior of even order neutral differential equations with variable coefficients. *Appl. Math. Lett.* **2006**, *19*, 1202–1206. [CrossRef]
33. Liu, L.; Bai, Y. New oscillation criteria for second-order nonlinear neutral delay differential equations. *J. Comput. Appl. Math.* **2009**, *231*, 657–663. [CrossRef]

 © 2020 by the authors. Licensee MDPI, Basel, Switzerland. This article is an open access article distributed under the terms and conditions of the Creative Commons Attribution (CC BY) license (http://creativecommons.org/licenses/by/4.0/).

Article

Initial Value Problem For Nonlinear Fractional Differential Equations With ψ-Caputo Derivative via Monotone Iterative Technique

Choukri Derbazi [1,†], Zidane Baitiche [1,†] and Mouffak Benchohra [2,†] and Alberto Cabada [3,*,†]

1. Laboratory of Mathematics and Applied Sciences, University of Ghardaia, Metlili 47000, Algeria; choukriedp@yahoo.com (C.D.); baitichezidane19@gmail.com (Z.B.)
2. Laboratory of Mathematics, Djillali Liabes University of Sidi-Bel-Abbes, Sidi Bel Abbès 22000, Algeria; benchohra@yahoo.com
3. Departamento de Estatística, Análise Matemática e Optimización, Facultade de Matemáticas, Instituto de Matemáticas, Universidade de Santiago de Compostela, 15705 Santiago de Compostela, Spain
* Correspondence: alberto.cabada@usc.gal
† These authors contributed equally to this work.

Received: 6 May 2020; Accepted: 16 May 2020; Published: 21 May 2020

Abstract: In this article, we discuss the existence and uniqueness of extremal solutions for nonlinear initial value problems of fractional differential equations involving the ψ-Caputo derivative. Moreover, some uniqueness results are obtained. Our results rely on the standard tools of functional analysis. More precisely we apply the monotone iterative technique combined with the method of upper and lower solutions to establish sufficient conditions for existence as well as the uniqueness of extremal solutions to the initial value problem. An illustrative example is presented to point out the applicability of our main results.

Keywords: ψ-Caputo fractional derivative; Cauchy problem extremal solutions; monotone iterative technique; upper and lower solutions

1. Introduction

Fractional differential equations have been applied in many fields of engineering, physics, biology, and chemistry see [1–4]. Moreover, to get a couple of developments about the theory of fractional differential equations, one can allude to the monographs of Abbas et al. [5–7], Kilbas et al. [8], Miller and Ross [9], Podlubny [10], and Zhou [11,12], as well as to the papers by Agarwal, et al. [13], Benchohra, et al. [14–16], and the references therein. In the recent past, Almeida in [17] presented a new fractional differentiation operator called by ψ-Caputo fractional operator. For more details see [18–23], and the references given therein.

At the present day, different kinds of fixed point theorems are widely used as fundamental tools in order to prove the existence and uniqueness of solutions for various classes of nonlinear fractional differential equations for details, we refer the reader to a series of papers [24–30] and the references therein, but here we focus on those using the monotone iterative technique, coupled with the method of upper and lower solutions. This method is a very useful tool for proving the existence and approximation of solutions to many applied problems of nonlinear differential equations and integral equations (see [31–42]). However, as far as we know, there is no work yet reported on the existence of extremal solutions for the Cauchy problem with ψ-Caputo fractional derivative. Motivated by this fact, in this paper we deal with the existence and uniqueness of extremal solutions for the following initial value problem of fractional differential equations involving the ψ-Caputo derivative:

$$\begin{cases} {}^cD_{a+}^{\alpha;\psi}x(t) = f(t,x(t)), \ t \in J := [a,b], \\ x(a) = a^*, \end{cases} \quad (1)$$

where ${}^cD_{a+}^{\alpha;\psi}$ is the ψ-Caputo fractional derivative of order $\alpha \in (0,1]$, $f: [a,b] \times \mathbb{R} \longrightarrow \mathbb{R}$ is a given continuous function and $a^* \in \mathbb{R}$.

The rest of the paper is organized as follows: in Section 2, we give some necessary definitions and lemmas. The main results are given in Section 3. Finally, an example is presented to illustrate the applicability of the results developed.

2. Preliminaries

In this section, we introduce some notations and definitions of fractional calculus and present preliminary results needed in our proofs later.

We begin by defining ψ-Riemann-Liouville fractional integrals and derivatives. In what follows,

Definition 1 ([8,17]). *For $\alpha > 0$, the left-sided ψ-Riemann-Liouville fractional integral of order α for an integrable function $x: J \longrightarrow \mathbb{R}$ with respect to another function $\psi: J \longrightarrow \mathbb{R}$ that is an increasing differentiable function such that $\psi'(t) \neq 0$, for all $t \in J$ is defined as follows*

$$I_{a+}^{\alpha;\psi}x(t) = \frac{1}{\Gamma(\alpha)} \int_a^t \psi'(s)(\psi(t) - \psi(s))^{\alpha-1}x(s)\,ds, \quad (2)$$

where Γ is the classical Euler Gamma function.

Definition 2 ([17]). *Let $n \in \mathbb{N}$ and let $\psi, x \in C^n(J, \mathbb{R})$ be two functions such that ψ is increasing and $\psi'(t) \neq 0$, for all $t \in J$. The left-sided ψ-Riemann–Liouville fractional derivative of a function x of order α is defined by*

$$\begin{aligned} D_{a+}^{\alpha;\psi}x(t) &= \left(\frac{1}{\psi'(t)}\frac{d}{dt}\right)^n I_{a+}^{n-\alpha;\psi}x(t) \\ &= \frac{1}{\Gamma(n-\alpha)}\left(\frac{1}{\psi'(t)}\frac{d}{dt}\right)^n \int_a^t \psi'(s)(\psi(t)-\psi(s))^{n-\alpha-1}x(s)\,ds, \end{aligned}$$

where $n = [\alpha] + 1$.

Definition 3 ([17]). *Let $n \in \mathbb{N}$ and let $\psi, x \in C^n(J, \mathbb{R})$ be two functions such that ψ is increasing and $\psi'(t) \neq 0$, for all $t \in J$. The left-sided ψ-Caputo fractional derivative of x of order α is defined by*

$${}^cD_{a+}^{\alpha;\psi}x(t) = I_{a+}^{n-\alpha;\psi}\left(\frac{1}{\psi'(t)}\frac{d}{dt}\right)^n x(t),$$

where $n = [\alpha] + 1$ for $\alpha \notin \mathbb{N}$, $n = \alpha$ for $\alpha \in \mathbb{N}$.

To simplify notation, we will use the abbreviated symbol

$$x_\psi^{[n]}(t) = \left(\frac{1}{\psi'(t)}\frac{d}{dt}\right)^n x(t).$$

From the definition, it is clear that

$$^cD_{a^+}^{\alpha;\psi}x(t) = \begin{cases} \int_a^t \frac{\psi'(s)(\psi(t)-\psi(s))^{n-\alpha-1}}{\Gamma(n-\alpha)} x_\psi^{[n]}(s)\mathrm{d}s, & \text{if } \alpha \notin \mathbb{N}, \\ x_\psi^{[n]}(t), & \text{if } \alpha \in \mathbb{N}. \end{cases} \quad (3)$$

We note that if $x \in C^n(J,\mathbb{R})$ the ψ-Caputo fractional derivative of order α of x is determined as

$$^cD_{a^+}^{\alpha;\psi}x(t) = D_{a^+}^{\alpha;\psi}\left[x(t) - \sum_{k=0}^{n-1} \frac{x_\psi^{[k]}(a)}{k!}(\psi(t)-\psi(a))^k\right].$$

(see, for instance, [17], Theorem 3).

Lemma 1 ([20]). *Let $\alpha, \beta > 0$, and $x \in L^1(J,\mathbb{R})$. Then*

$$I_{a^+}^{\alpha;\psi} I_{a^+}^{\beta;\psi} x(t) = I_{a^+}^{\alpha+\beta;\psi} x(t), \text{ a.e. } t \in J.$$

In particular, if $x \in C(J,\mathbb{R})$, then $I_{a^+}^{\alpha;\psi} I_{a^+}^{\beta;\psi} x(t) = I_{a^+}^{\alpha+\beta;\psi} x(t), t \in J$.

Lemma 2 ([20]). *Let $\alpha > 0$, The following holds:*
If $x \in C(J,\mathbb{R})$ then

$$^cD_{a^+}^{\alpha;\psi} I_{a^+}^{\alpha;\psi} x(t) = x(t), \ t \in J.$$

If $x \in C^n(J,\mathbb{R})$, $n-1 < \alpha < n$. Then

$$I_{a^+}^{\alpha;\psi}\ ^cD_{a^+}^{\alpha;\psi} x(t) = x(t) - \sum_{k=0}^{n-1} \frac{x_\psi^{[k]}(a)}{k!} [\psi(t)-\psi(a)]^k, \quad t \in J.$$

Lemma 3 ([8,20]). *Let $t > a$, $\alpha \geq 0$, and $\beta > 0$. Then*

- $I_{a^+}^{\alpha;\psi}(\psi(t)-\psi(a))^{\beta-1} = \frac{\Gamma(\beta)}{\Gamma(\beta+\alpha)}(\psi(t)-\psi(a))^{\beta+\alpha-1}$,
- $^cD_{a^+}^{\alpha;\psi}(\psi(t)-\psi(a))^{\beta-1} = \frac{\Gamma(\beta)}{\Gamma(\beta-\alpha)}(\psi(t)-\psi(a))^{\beta-\alpha-1}$,
- $^cD_{a^+}^{\alpha;\psi}(\psi(t)-\psi(a))^k = 0$, for all $k \in \{0,\ldots,n-1\}, n \in \mathbb{N}$.

Definition 4 ([43]). *The one-parameter Mittag–Leffler function $\mathbb{E}_\alpha(\cdot)$, is defined as:*

$$\mathbb{E}_\alpha(z) = \sum_{k=0}^\infty \frac{z^k}{\Gamma(\alpha k+1)}, \quad (z \in \mathbb{R}, \alpha > 0).$$

Definition 5 ([43]). *The Two-parameter Mittag–Leffler function $\mathbb{E}_{\alpha,\beta}(\cdot)$, is defined as:*

$$\mathbb{E}_{\alpha,\beta}(z) = \sum_{k=0}^\infty \frac{z^k}{\Gamma(\alpha k+\beta)}, \ \alpha,\beta > 0 \text{ and } z \in \mathbb{R}. \quad (4)$$

Theorem 1 (Weissinger's fixed point theorem [44]). *Assume (E,d) to be a non empty complete metric space and let $\beta_j \geq 0$ for every $j \in \mathbb{N}$ such that $\sum_{j=0}^{n-1} \beta_j$ converges. Furthermore, let the mapping $\mathbb{T}: E \to E$ satisfy the inequality*

$$d(\mathbb{T}^j u, \mathbb{T}^j v) \leq \beta_j d(u,v),$$

for every $j \in \mathbb{N}$ and every $u, v \in E$. Then, $\mathbb{T}$ has a unique fixed point u^*. Moreover, for any $v_0 \in E$, the sequence $\{\mathbb{T}^j v_0\}_{j=1}^\infty$ converges to this fixed point u^*.

3. Main Results

Let us recall the definition and lemma of a solution for problem (1). First of all, we define what we mean by a solution for the boundary value problem (1).

Definition 6. *A function $x \in C(J, \mathbb{R})$ is said to be a solution of Equation (1) if x satisfies the equation ${}^c D_{a+}^{\alpha;\psi} x(t) = f(t, x(t))$, for each $t \in J$ and the condition*

$$x(a) = a^*.$$

For the existence of solutions for problem (1) we need the following lemma for a general linear equation of $\alpha > 0$, that generalizes expression (3.1.34) in [8].

Lemma 4. *For a given $h \in C(J, \mathbb{R})$ and $\alpha \in (n-1, n]$, with $n \in \mathbb{N}$, the linear fractional initial value problem*

$$\begin{cases} {}^c D_{a+}^{\alpha;\psi} x(t) + r x(t) = h(t), & t \in J := [a, b], \\ x_\psi^{[k]}(a) = a_k, & k = 0, \ldots, n-1, \end{cases} \quad (5)$$

has a unique solution given by

$$\begin{aligned}
x(t) &= \sum_{k=0}^{n-1} \frac{a_k}{k!} [\psi(t) - \psi(a)]^k - r \mathcal{I}_{a+}^{\alpha;\psi} x(t) + \mathcal{I}_{a+}^{\alpha;\psi} h(t) \\
&= \sum_{k=0}^{n-1} \frac{a_k}{k!} [\psi(t) - \psi(a)]^k - \frac{r}{\Gamma(\alpha)} \int_a^t \psi'(s)(\psi(t) - \psi(s))^{\alpha-1} x(s) ds \\
&\quad + \frac{1}{\Gamma(\alpha)} \int_a^t \psi'(s)(\psi(t) - \psi(s))^{\alpha-1} h(s) ds.
\end{aligned} \quad (6)$$

Moreover, the explicit solution of the Volterra integral equation (6) can be represented by

$$\begin{aligned}
x(t) &= \sum_{k=0}^{n-1} a_k [\psi(t) - \psi(a)]^k \mathbb{E}_{\alpha, k+1}\left(-r(\psi(t) - \psi(a))^\alpha\right) \\
&\quad + \int_a^t \psi'(s)(\psi(t) - \psi(s))^{\alpha-1} \mathbb{E}_{\alpha, \alpha}\left(-r(\psi(t) - \psi(a))^\alpha\right) h(s) ds,
\end{aligned} \quad (7)$$

where $\mathbb{E}_{\alpha, \beta}(\cdot)$ is the two-parametric Mittag–Leffler function defined in (4).

Proof. Since $\alpha \in (n-1, n]$, from Lemma 2 we know that the Cauchy problem (5) is equivalent to the following Volterra integral equation

$$\begin{aligned}
x(t) &= \sum_{k=0}^{n-1} \frac{a_k}{k!} [\psi(t) - \psi(a)]^k - r \mathcal{I}_{a+}^{\alpha;\psi} x(t) + \mathcal{I}_{a+}^{\alpha;\psi} h(t) \\
&= \sum_{k=0}^{n-1} \frac{a_k}{k!} [\psi(t) - \psi(a)]^k - \frac{r}{\Gamma(\alpha)} \int_a^t \psi'(s)(\psi(t) - \psi(s))^{\alpha-1} x(s) ds \\
&\quad + \frac{1}{\Gamma(\alpha)} \int_a^t \psi'(s)(\psi(t) - \psi(s))^{\alpha-1} h(s) ds.
\end{aligned}$$

Note that the above equation can be written in the following form

$$x(t) = \mathcal{T} x(t),$$

where the operator $\mathcal{T}$ is defined by

$$\mathcal{T}x(t) = \sum_{k=0}^{n-1} \frac{a_k}{k!} [\psi(t) - \psi(a)]^k - r\mathcal{I}_{a+}^{\alpha;\psi} x(t) + \mathcal{I}_{a+}^{\alpha;\psi} h(t).$$

Let $n \in \mathbb{N}$ and $x, y \in C(J, \mathbb{R})$. Then, we have

$$\begin{aligned}
|\mathcal{T}^n(x)(t) - \mathcal{T}^n(y)(t)| &= \left| -r\mathcal{I}_{a+}^{\alpha;\psi} \left(\mathcal{T}^{n-1}x(t) - \mathcal{T}^{n-1}y(t) \right) \right| \\
&= \left| -r\mathcal{I}_{a+}^{\alpha;\psi} \left(-r\mathcal{I}_{a+}^{\alpha;\psi} \left(\mathcal{T}^{n-2}x(t) - \mathcal{T}^{n-2}y(t) \right) \right) \right| \\
&\vdots \\
&= \left| (-r)^n \mathcal{I}_{a+}^{n\alpha;\psi} (x(t) - y(t)) \right| \\
&\leq \frac{(r(\psi(b) - \psi(a))^\alpha)^n}{\Gamma(n\alpha + 1)} \|x - y\|.
\end{aligned}$$

Hence, we have

$$\|\mathcal{T}^n(x) - \mathcal{T}^n(y)\| \leq \frac{r^n (\psi(b) - \psi(a))^{n\alpha}}{\Gamma(n\alpha + 1)} \|x - y\|.$$

It's well known that

$$\sum_{n=0}^{\infty} \frac{r^n (\psi(b) - \psi(a))^{n\alpha}}{\Gamma(n\alpha + 1)} = \mathbb{E}_\alpha \left(r(\psi(b) - \psi(a))^\alpha \right),$$

it follows that the mapping $\mathcal{T}^n$ is a contraction. Hence, by Weissinger's fixed point theorem, $\mathcal{T}$ has a unique fixed point. That is (5) has a unique solution.

Now we apply the method of successive approximations to prove that the integral Equation (6) can be expressed by

$$\begin{aligned}
x(t) = &\sum_{k=0}^{n-1} a_k [\psi(t) - \psi(a)]^k \mathbb{E}_{\alpha, k+1} \left(-r(\psi(t) - \psi(a))^\alpha \right) \\
&+ \int_a^t \psi'(s) (\psi(t) - \psi(s))^{\alpha-1} \mathbb{E}_{\alpha, \alpha} \left(-r(\psi(t) - \psi(a))^{\alpha-1} \right) h(s) ds.
\end{aligned}$$

For this, we set

$$\begin{cases}
x_0(t) = \sum_{k=0}^{n-1} \frac{a_k}{k!} [\psi(t) - \psi(a)]^k \\
x_m(t) = x_0(t) - \frac{r}{\Gamma(\alpha)} \int_a^t \psi'(s) (\psi(t) - \psi(s))^{\alpha-1} x_{m-1}(s) ds \\
\qquad + \frac{1}{\Gamma(\alpha)} \int_a^t \psi'(s) (\psi(t) - \psi(s))^{\alpha-1} h(s) ds.
\end{cases} \quad (8)$$

It follows from Equation (8) and Lemma 3 that

$$\begin{aligned}
x_1(t) &= x_0(t) - r\mathcal{I}_{a+}^{\alpha;\psi} x_0(t) + \mathcal{I}_{a+}^{\alpha;\psi} h(t) \\
&= \sum_{k=0}^{n-1} \frac{a_k}{k!} [\psi(t) - \psi(a)]^k - r \sum_{k=0}^{n-1} \frac{a_k}{\Gamma(\alpha + k + 1)} [\psi(t) - \psi(a)]^{\alpha+k} + \mathcal{I}_{a+}^{\alpha;\psi} h(t).
\end{aligned} \quad (9)$$

Similarly, Equations (8) and (9) and Lemmas 1 and 3 yield

$$x_2(t) = x_0(t) - r\mathcal{I}_{a^+}^{\alpha;\psi} x_1(t) + \mathcal{I}_{a^+}^{\alpha;\psi} h(t)$$

$$= \sum_{k=0}^{n-1} \frac{a_k}{k!} [\psi(t) - \psi(a)]^k - r\mathcal{I}_{a^+}^{\alpha;\psi} \left(\sum_{k=0}^{n-1} \frac{a_k}{k!} [\psi(t) - \psi(a)]^k \right.$$

$$\left. - r \sum_{k=0}^{n-1} \frac{a_k}{\Gamma(\alpha + k + 1)} [\psi(t) - \psi(a)]^{\alpha+k} + \mathcal{I}_{a^+}^{\alpha;\psi} h(t) \right) + \mathcal{I}_{a^+}^{\alpha;\psi} h(t)$$

$$= \sum_{k=0}^{n-1} \frac{a_k}{k!} [\psi(t) - \psi(a)]^k - r \sum_{k=0}^{n-1} \frac{a_k}{\Gamma(\alpha + k + 1)} [\psi(t) - \psi(a)]^{\alpha+k}$$

$$+ r^2 \sum_{k=0}^{n-1} \frac{a_k}{\Gamma(2\alpha + k + 1)} [\psi(t) - \psi(a)]^{2\alpha+k} - r\mathcal{I}_{a^+}^{2\alpha;\psi} h(t) + \mathcal{I}_{a^+}^{\alpha;\psi} h(t)$$

$$= \sum_{l=0}^{2} \sum_{k=0}^{n-1} \frac{(-r)^l a_k}{\Gamma(l\alpha + k + 1)} [\psi(t) - \psi(a)]^{l\alpha+k} + \int_a^t \psi'(s) \sum_{l=0}^{1} \frac{(-r)^{l-1} (\psi(t) - \psi(s))^{l\alpha+\alpha-1}}{\Gamma(l\alpha + \alpha)} h(s) ds.$$

Continuing this process, we derive the following relation

$$x_m(t) = \sum_{l=0}^{m} \sum_{k=0}^{n-1} \frac{(-r)^l a_k}{\Gamma(l\alpha + k + 1)} [\psi(t) - \psi(a)]^{l\alpha+k} + \int_a^t \psi'(s) \sum_{l=0}^{m-1} \frac{(-r)^{l-1} (\psi(t) - \psi(s))^{l\alpha+\alpha-1}}{\Gamma(l\alpha + \alpha)} h(s) ds.$$

Taking the limit as $n \to \infty$, we obtain the following explicit solution $x(t)$ to the integral Equation (6):

$$x(t) = \sum_{l=0}^{\infty} \sum_{k=0}^{n-1} \frac{(-r)^l a_k}{\Gamma(l\alpha + k + 1)} [\psi(t) - \psi(a)]^{l\alpha+k} + \int_a^t \psi'(s) \sum_{l=0}^{\infty} \frac{(-r)^{l-1} (\psi(t) - \psi(s))^{l\alpha+\alpha-1}}{\Gamma(l\alpha + \alpha)} h(s) ds$$

$$= \sum_{k=0}^{n-1} a_k (\psi(t) - \psi(a))^k \sum_{l=0}^{\infty} \frac{(-r)^l}{\Gamma(l\alpha + k + 1)} [\psi(t) - \psi(a)]^{l\alpha}$$

$$+ \int_a^t \psi'(s) (\psi(t) - \psi(s))^{\alpha-1} \sum_{l=0}^{\infty} \frac{(-r)^{l-1} (\psi(t) - \psi(s))^{l\alpha}}{\Gamma(l\alpha + \alpha)} h(s) ds.$$

Taking into account (4), we get

$$x(t) = \sum_{k=0}^{n-1} a_k [\psi(t) - \psi(a)]^k \mathbb{E}_{\alpha, k+1} \left(-r(\psi(t) - \psi(a))^\alpha \right)$$

$$+ \int_a^t \psi'(s) (\psi(t) - \psi(s))^{\alpha-1} \mathbb{E}_{\alpha, \alpha} \left(-r(\psi(t) - \psi(s))^\alpha \right) h(s) ds.$$

Then the proof is completed. □

Lemma 5 (Comparison result). *Let $\alpha \in (0, 1]$ be fixed and $r \in \mathbb{R}$. If $\rho \in C(J, \mathbb{R})$ satisfies the following inequalities*

$$\begin{cases} {}^c D_{a^+}^{\alpha;\psi} \rho(t) \geq -r\rho(t), & t \in [a, b], \\ \rho(a) \geq 0, \end{cases} \tag{10}$$

then $\rho(t) \geq 0$ for all $t \in J$.

Proof. Using the integral representation (7) and the fact that, $\mathbb{E}_{\alpha,1}(z) \geq 0$ and $\mathbb{E}_{\alpha,\alpha}(z) \geq 0$ for all $\alpha \in (0, 1]$ and $z \in \mathbb{R}$, (see [45]) it suffices to take $h(t) = {}^c \mathcal{D}_{a^+}^{\alpha;\psi} \rho(t) + r\rho(t) \geq 0$ with initial conditions $\rho(a) = a^* \geq 0$. □

Definition 7. *A function $x_0 \in C(J, \mathbb{R})$ is said to be a lower solution of the problem (1), if it satisfies*

$$\begin{cases} ^cD_{a^+}^{\alpha;\psi} x_0(t) \leq f(t, x_0), & t \in (a, b], \\ x_0(a) \leq a^*. \end{cases} \quad (11)$$

Definition 8. *A function $y_0 \in C(J, \mathbb{R})$ is called an upper solution of problem (1), if it satisfies*

$$\begin{cases} ^cD_{a^+}^{\alpha;\psi} y_0(t) \geq f(t, y_0), & t \in (a, b], \\ y_0(a) \geq a^*. \end{cases} \quad (12)$$

Theorem 2. *Let the function $f \in C(J \times \mathbb{R}, \mathbb{R})$. In addition assume that:*

(H_1) *There exist $x_0, y_0 \in C(J, \mathbb{R})$ such that x_0 and y_0 are lower and upper solutions of problem (1), respectively, with $x_0(t) \leq y_0(t), t \in J$.*

(H_2) *There exists a constant $r \in \mathbb{R}$ such that*

$$f(t, y) - f(t, x) \geq -r(y - x) \quad \text{for } x_0 \leq x \leq y \leq y_0.$$

Then there exist monotone iterative sequences $\{x_n\}$ and $\{y_n\}$, which converge uniformly on the interval J to the extremal solutions of (1) in the sector $[x_0, y_0]$, where

$$[x_0, y_0] = \{z \in C(J, \mathbb{R}) : x_0(t) \leq z(t) \leq y_0(t), \quad t \in J\}.$$

Proof. First, for any $x_0(t), y_0(t) \in C(J, \mathbb{R})$, we consider the following linear initial value problems of fractional order:

$$\begin{cases} ^cD_{a^+}^{\alpha;\psi} x_{n+1}(t) = f(t, x_n(t)) - r(x_{n+1}(t) - x_n(t)), & t \in J, \\ x_{n+1}(a) = a^*, \end{cases} \quad (13)$$

and

$$\begin{cases} ^cD_{a^+}^{\alpha;\psi} y_{n+1}(t) = f(t, y_n(t)) - r(y_{n+1}(t) - y_n(t)), & t \in J, \\ y_{n+1}(a) = a^*. \end{cases} \quad (14)$$

By Lemma 4, we know that (13) and (14) have unique solutions in $C(J, \mathbb{R})$ which are defined as follows:

$$x_{n+1}(t) = a^* \mathbb{E}_{\alpha,1}\left(-r(\psi(t) - \psi(a))^\alpha\right) \\ + \int_a^t \psi'(s)(\psi(t) - \psi(s))^{\alpha-1} \mathbb{E}_{\alpha,\alpha}\left(-r(\psi(t) - \psi(s))^\alpha\right)\left(f(s, x_n(s)) + rx_n(s)\right)ds, \ t \in J, \quad (15)$$

$$y_{n+1}(t) = a^* \mathbb{E}_{\alpha,\alpha}\left(-r(\psi(t) - \psi(a))^\alpha\right) \\ + \int_a^t \psi'(s)(\psi(t) - \psi(s))^{\alpha-1} \mathbb{E}_{\alpha,\alpha}\left(-r(\psi(t) - \psi(s))^\alpha\right)\left(f(s, y_n(s)) + ry_n(s)\right)ds, \ t \in J. \quad (16)$$

We will divide the proof into three steps.

Step 1: We show that the sequences $x_n(t), y_n(t)(n \geq 1)$ are lower and upper solutions of problem (1), respectively and the following relation holds

$$x_0(t) \leq x_1(t) \leq \cdots \leq x_n(t) \leq \cdots \leq y_n(t) \leq \cdots \leq y_1(t) \leq y_0(t), \quad t \in J. \quad (17)$$

First, we prove that

$$x_0(t) \leq x_1(t) \leq y_1(t) \leq y_0(t), \quad t \in J. \tag{18}$$

Set $\rho(t) = x_1(t) - x_0(t)$. From (13) and Definition 7, we obtain

$$\begin{aligned} {}^c D_{a+}^{\alpha;\psi}\rho(t) &= {}^c D_{a+}^{\alpha;\psi} x_1(t) - {}^c D_{a+}^{\alpha;\psi} x_0(t) \\ &\geq f(t, x_0(t)) - r(x_1(t) - x_0(t)) - f(t, x_0(t)) \\ &= -r\rho(t). \end{aligned}$$

Again, since

$$\rho(a) = x_1(a) - x_0(a) = a^* - x_0(a) \geq 0.$$

By Lemma 5, $\rho(t) \geq 0$, for $t \in J$. That is, $x_0(t) \leq x_1(t)$. Similarly, we can show that $y_1(t) \leq y_0(t)$, $t \in J$.

Now, let $\rho(t) = y_1(t) - x_1(t)$. From (13), (14) and (H2), we get

$$\begin{aligned} {}^c D_{a+}^{\alpha;\psi}\rho(t) &= {}^c D_{a+}^{\alpha;\psi} y_1(t) - {}^c D_{a+}^{\alpha;\psi} x_1(t) \\ &= f(t, y_0(t)) - r(y_1(t) - y_0(t)) - f(t, x_0(t)) + r(x_1(t) - x_0(t)) \\ &= f(t, y_0(t)) - f(t, x_0(t)) - r(y_1(t) - y_0(t)) + r(x_1(t) - x_0(t)) \\ &\geq -r(y_0(t) - x_0(t)) - r(y_1(t) - y_0(t)) + r(x_1(t) - x_0(t)) \\ &= -r\rho(t). \end{aligned}$$

Since, $\rho(a) = x_1(a) - y_1(a) = a^* - a^* = 0$. By Lemma 5, we get $x_1(t) \leq y_1(t)$, $t \in J$.

Secondly, we show that $x_1(t), y_1(t)$ are lower and upper solutions of problem (1), respectively. Since x_0 and y_0 are lower and upper solutions of problem (1), by (H2), it follows that

$${}^c D_{a+}^{\alpha;\psi} x_1(t) = f(t, x_0(t)) - r(x_1(t) - x_0(t)) \leq f(t, x_1(t)),$$

also $x_1(a) = a^*$. Therefore, $x_1(t)$ is a lower solution of problem (1). Similarly, it can be obtained that $y_1(t)$ is an upper solution of problem (1).

By the above arguments and mathematical induction, we can show that the sequences $x_n(t), y_n(t), (n \geq 1)$ are lower and upper solutions of problem (1), respectively and the following relation holds

$$x_0(t) \leq x_1(t) \leq \cdots \leq x_n(t) \leq \cdots \leq y_n(t) \leq \cdots \leq y_1(t) \leq y_0(t), \quad t \in J.$$

Step 2: The sequences $\{x_n(t)\}$, $\{y_n(t)\}$ converge uniformly to their limit functions $x^*(t), y^*(t)$, respectively.

Note that the sequence $\{x_n(t)\}$ is monotone nondecreasing and is bounded from above by $y_0(t)$. Since the sequence $\{y_n(t)\}$ is monotone nonincreasing and is bounded from below by $x_0(t)$, therefore the pointwise limits exist and these limits are denoted by x^* and y^*. Moreover, since $\{x_n(t)\}$, $\{y_n(t)\}$ are sequences of continuous functions defined on the compact set $[a, b]$, hence by Dini's theorem [46], the convergence is uniform. This is

$$\lim_{n \to \infty} x_n(t) = x^*(t) \quad \text{and} \quad \lim_{n \to \infty} y_n(t) = y^*(t),$$

uniformly on $t \in J$ and the limit functions x^*, y^* satisfy problem (1). Furthermore, x^* and y^* satisfy the relation

$$x_0 \leq x_1 \leq \cdots \leq x_n \leq x^* \leq y^* \leq \cdots \leq y_n \leq \cdots \leq y_1 \leq y_0.$$

Step 3: We prove that x^* and y^* are extremal solutions of problem (1) in $[x_0, y_0]$.

Let $z \in [x_0, y_0]$ be any solution of (1). We assume that the following relation holds for some $n \in \mathbb{N}$:

$$x_n(t) \leq z(t) \leq y_n(t), \quad t \in J. \tag{19}$$

Let $\rho(t) = z(t) - x_{n+1}(t)$. We have

$$\begin{aligned} {}^c D_{a+}^{\alpha;\psi} \rho(t) &= {}^c D_{a+}^{\alpha;\psi} z(t) - {}^c D_{a+}^{\alpha;\psi} x_{n+1}(t) \\ &= f(t, z(t)) - f(t, x_n(t)) + r(x_{n+1}(t) - x_n(t)) \\ &\geq -r(z(t) - x_n(t)) + r(x_{n+1}(t) - x_n(t)) \\ &= -r\rho(t). \end{aligned}$$

Furthermore, $\rho(a) = z(a) - x_{n+1}(a) = a^* - a^* = 0$. By Lemma 5, we obtain $\rho(t) \geq 0$, $t \in J$, which means

$$x_{n+1}(t) \leq z(t), \ t \in J.$$

Using the same method, we can show that

$$z(t) \leq y_{n+1}(t), \ t \in J.$$

Hence, we have

$$x_{n+1}(t) \leq z(t) \leq y_{n+1}(t), \ t \in J.$$

Therefore, (19) holds on J for all $n \in \mathbb{N}$. Taking the limit as $n \to \infty$ on both sides of (19), we get

$$x^* \leq z \leq y^*.$$

Therefore x^*, y^* are the extremal solutions of (1) in $[x_0, y_0]$. This completes the proof. □

Now, we shall prove the uniqueness of the solution of the system (1) by monotone iterative technique.

Theorem 3. *Suppose that (H1) and (H2) are satisfied. Furthermore, we impose that:*

(H3) There exists a constant $r^* \geq -r$ such that

$$f(t, y) - f(t, x) \leq r^*(y - x),$$

for every $x_0 \leq x \leq y \leq y_0$, $t \in J$. Then problem (1) has a unique solution between x_0 and y_0.

Proof. From the Theorem 2, we know that $x^*(t)$ and $y^*(t)$ are the extremal solutions of the IVP (1) and $x^*(t) \leq y^*(t)$, $t \in J$. It is sufficient to prove $x^*(t) \geq y^*(t)$, $t \in J$. In fact, let $\rho(t) = x^*(t) - y^*(t)$, $t \in J$, in view of (H3), we have

$$\begin{aligned} {}^c D_{a+}^{\alpha;\psi} \rho(t) &= {}^c D_{a+}^{\alpha;\psi} x^*(t) - {}^c D_{a+}^{\alpha;\psi} y^*(t) \\ &= f(t, x^*(t)) - f(t, y^*(t)) \\ &\geq r^*(x^*(t) - y^*(t)) \\ &= r^* \rho(t). \end{aligned}$$

Furthermore, $\rho(a) = x^*(a) - y^*(a) = a^* - a^* = 0$. From Lemma 5, it follows that $\rho(t) \geq 0$, $t \in J$. Hence, we obtain

$$x^*(t) \geq y^*(t), \quad t \in J.$$

Therefore, $x^* \equiv y^*$ is the unique solution of the Cauchy problem (1) in $[x_0, y_0]$. This ends the proof of Theorem 3. □

As a direct consequence of the previous result, we arrive at the following one

Corollary 1. *Suppose that (H1) is satisfied and that $f \in C(E, \mathbb{R})$, is differentiable with respect to x and $\partial f / \partial x \in C(E, \mathbb{R})$, with*

$$E = \{(t, x) \in \mathbb{R}^2, \quad \text{such that} \quad x_0(t) \le x \le y_0(t)\}.$$

Then problem (1) has a unique solution between x_0 and y_0.

Proof. The proof follows immediately from the fact that E is a compact set and, as a consequence, $\partial f / \partial x$ is bounded in E. □

4. An Example

Example 1. *Consider the following problem:*

$$\begin{cases} \mathcal{D}_{0^+}^{\frac{1}{2}} x(t) = 1 - x^2(t) + 2t, \quad t \in J := [0, 1], \\ x(0) = 1 \end{cases} \tag{20}$$

Note that, this problem is a particular case of IVP (1), where

$$\alpha = \frac{1}{2}, \; a = 0, \; b = 1, \; a^* = 1, \; \psi(t) = t,$$

and $f : J \times \mathbb{R} \longrightarrow \mathbb{R}$ given by

$$f(t, x) = 1 - x^2 + 2t, \quad \text{for } t \in J, x \in \mathbb{R}.$$

Taking $x_0(t) \equiv 0$ and $y_0(t) = 1 + t$, it is not difficult to verify that x_0, y_0 are lower and upper solutions of (20), respectively, and $x_0 \le y_0$. So (H_1) of Theorem 2 holds
On the other hand, it is clear that the function f is continuous and satisfies

$$\left| \frac{f(t, x)}{\partial x}(t, x) \right| = |-2x| \le 4 \quad \text{for all } t \in [0, 1] \text{ and } 0 \le x \le t + 1.$$

Hence, by Corollary 1, the initial value problem (20) has a unique solution u^* and there exist monotone iterative sequences $\{x_n\}$ and $\{y_n\}$ converging uniformly to u^*. Furthermore, we have the following iterative sequences

$$x_{n+1}(t) = \mathbb{E}_{\frac{1}{2},1}(-4\sqrt{t}) + \int_0^t (t-s)^{-1/2} \mathbb{E}_{\frac{1}{2},\frac{1}{2}}(-4\sqrt{t-s})(1 - x_n^2(s) + 2s + 4x_n(s)) ds, \; t \in J,$$

$$y_{n+1}(t) = \mathbb{E}_{\frac{1}{2},1}(-4\sqrt{t}) + \int_0^t (t-s)^{-1/2} \mathbb{E}_{\frac{1}{2},\frac{1}{2}}(-4\sqrt{t-s})(1 - y_n^2(s) + 2s + 4y_n(s)) ds, \; t \in J.$$

We notice that the sequences are obtained by solving a recurrence formula of the type $v_{n+1} = A v_n$, with A a suitable integral operator and v_0 given. So, by a simple numerical procedure, it is not difficult to represent some iterates of the recurrence sequence. We plot in Figure 1 the four first iterates of each sequence. We point out that the unique solution is lying within x_3 and y_3 which gives us a good approximation of such a solution.

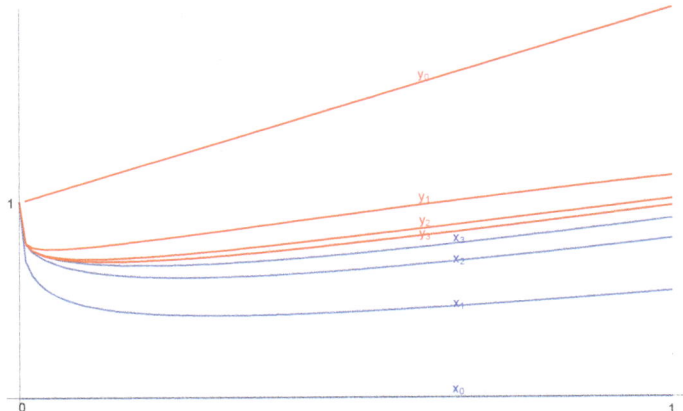

Figure 1. First four iterates for problem (20).

5. Conclusions

In previous sections, we have presented the existence and uniqueness of extremal solutions to a Cauchy problem with ψ-Caputo fractional derivative. Moreover, some uniqueness results are obtained. The proof of the existence results is based on the monotone iterative technique combined with the method of upper and lower solutions. Moreover, an example is presented to illustrate the validity of our main results. Our results are not only new in the given configuration but also correspond to some new situations associated with the specific values of the parameters involved in the given problem.

Author Contributions: conceptualization, C.D., Z.B., M.B. and A.C.; methodology, C.D., Z.B., M.B. and A.C.; formal analysis, C.D., Z.B., M.B. and A.C.; investigation, C.D., Z.B., M.B. and A.C.; writing—original draft preparation, C.D., Z.B., M.B. and A.C.; writing—review and editing, C.D., Z.B., M.B. and A.C.; funding acquisition, A.C. All authors have read and agreed to the published version of the manuscript.

Funding: The fourth author is supported by the Agencia Estatal de Investigación (AEI) of Spain under grant MTM2016-75140-P, co-financed by the European Community fund FEDER. The fourth author is also supported by Xunta de Galicia, project ED431C 2019/02 (Spain).

Conflicts of Interest: The authors declare no conflict of interest.

References

1. Hilfer, R. *Applications of Fractional Calculus in Physics*; World Scientific: Singapore, 2000.
2. Oldham, K.B. Fractional differential equations in electrochemistry. *Adv. Eng. Softw.* **2010**, *41*, 9–12. [CrossRef]
3. Sabatier, J.; Agrawal, O.P.; Machado, J.A.T. *Advances in Fractional Calculus-Theoretical Developments and Applications in Physics and Engineering*; Springer: Dordrecht, The Netherlands, 2007.
4. Tarasov, V.E. *Fractional Dynamics: Application of Fractional Calculus to Dynamics of Particles, Fields and Media*; Springer Science & Business Media: Berlin/Heidelberg, Germany, 2010.
5. Abbas, S.; Benchohra, M.; N'Guérékata, G.M. *Topics in Fractional Differential Equations*; Springer: New York, NY, USA, 2015.
6. Abbas, S.; Benchohra, M.; Graef, J.R.; Henderson, J. *Implicit Fractional Differential and Integral Equations: Existence and Stability*; De Gruyter: Berlin, Germany, 2018.
7. Abbas, S.; Benchohra, M.; N'Guérékata, G.M. *Advanced Fractional Differential and Integral Equations*; Nova Sci. Publ.: New York, NY, USA, 2014.
8. Kilbas, A.A.; Srivastava, H.M.; Trujillo, J.J. *Theory and Applications of Fractional Differential Equations*; vol. 204 of North-Holland Mathematics Sudies; Elsevier Science B.V.: Amsterdam, The Netherlands, 2006.
9. Miller, K.S.; Ross, B. *An Introduction to Fractional Calculus and Fractional Differential Equations*; Wiley: New York, NY, USA, 1993.
10. Podlubny, I. *Fractional Differential Equations*; Academic Press: San Diego, CA, USA, 1999.

11. Zhou, Y. *Basic Theory of Fractional Differential Equations*; World Scientific: Singapore, 2014.
12. Zhou, Y. *Fractional Evolution Equations and Inclusions*; Analysis and Control; Elsevier, Acad. Press: Cambridge, MA, USA, 2016.
13. Agarwal, R.P.; Benchohra, M.; Hamani, S. A survey onexistence results for boundary value problems of nonlinear fractional differential equations and inclusions. *Acta Appl. Math.* **2010**, *109*, 973–1033. [CrossRef]
14. Benchohra, M.; Graef, J.R.; Hamani, S. Existence results for boundary value problems with non-linear fractional differential equations. *Appl. Anal.* **2008**, *87*, 851–863. [CrossRef]
15. Benchohra, M.; Hamani, S.; Ntouyas, S.K. Boundary value problems for differential equations with fractional order and nonlocal conditions. *Nonlinear Anal.* **2009**, *71*, 2391–2396. [CrossRef]
16. Benchohra, M.; Lazreg, J.E. Existence results for nonlinear implicit fractional differential equations. *Surv. Math. Appl.* **2014**, *9*, 79–92.
17. Almeida, R. A Caputo fractional derivative of a function with respect to another function. *Commun. Nonlinear Sci.* **2017**, *44*, 460–481. [CrossRef]
18. Abdo, M.S.; Panchal, S.K.; Saeed, A.M. Fractional boundary value problem with ψ-Caputo fractional derivative. *Proc. Math. Sci.* **2019**, *129*, 14. [CrossRef]
19. Almeida, R. Fractional Differential Equations with Mixed Boundary Conditions. *Bull. Malays. Math. Sci. Soc.* **2019**, *42*, 1687–1697. [CrossRef]
20. Almeida, R.; Malinowska, A.B.; Monteiro, M.T.T. Fractional differential equations with a Caputo derivative with respect to a kernel function and their applications. *Math. Meth. Appl. Sci.* **2018**, *41*, 336–352. [CrossRef]
21. Almeida, R.; Malinowska, A.B.; Odzijewicz, T. Optimal Leader-Follower Control for the Fractional Opinion Formation Model. *J. Optim. Theory Appl.* **2019**, *182*, 1171–1185. [CrossRef]
22. Almeida, R.; Jleli, M.; Samet, B. A numerical study of fractional relaxation-oscillation equations involving ψ-Caputo fractional derivative. *Rev. R. Acad. Cienc. Exactas Fís. Nat. Ser. A Mat. RACSAM* **2019**, *113*, 1873–1891. [CrossRef]
23. Samet, B.; Aydi, H. Lyapunov-type inequalities for an anti-periodic fractional boundary value problem involving ψ-Caputo fractional derivative. *J. Inequal. Appl.* **2018**, *2018*, 286. [CrossRef] [PubMed]
24. Abbas, S.; Benchohra, M.; Samet, B.; Zhou, Y. Coupled implicit Caputo fractional q-difference systems. *Adv. Diff. Equ.* **2019**, *2019*, 527. [CrossRef]
25. Abbas, S.; Benchohra, M.; Hamidi, N.; Henderson, J. Caputo–Hadamard fractional differential equations in Banach spaces. *Fract. Calc. Appl. Anal.* **2018**, *21*, 1027–1045. [CrossRef]
26. Abbas, S.; Benchohra, M.; Hamani, S.; Henderson, J. Upper and lower solutions method for Caputo-Hadamard fractional differential inclusions. *Math. Morav.* **2019**, *23*, 107–118. [CrossRef]
27. Aghajani, A.; Pourhadi, E.; Trujillo, J.J. Application of measure of noncompactness to a Cauchy problem for fractional differential equations in Banach spaces. *Fract. Calc. Appl. Anal.* **2013**, *16*, 962–977. [CrossRef]
28. Kucche, K.D.; Mali, A.D.; Sousa, J.V.C. On the nonlinear Ψ-Hilfer fractional differential equations. *Comput. Appl. Math.* **2019**, *38*, 25. [CrossRef]
29. Wu, G.C.; Zeng, D.Q.; Baleanu, D. Fractional impulsive differential equations: Exact solutions, integral equations and short memory case. *Fract. Calc Appl. Anal.* **2019**, *22*, 180–192. [CrossRef]
30. Wu, G.C.; Deng, Z.G.; Baleanu, D.; Zeng, D.Q. New variable order fractional chaotic systems for fast image encryption. *Chaos* **2019**, *29*, 11. [CrossRef]
31. Ali, S.; Shah, K.; Jarad, F. On stable iterative solutions for a class of boundary value problem of nonlinear fractional order differential equations. *Math. Methods Appl. Sci.* **2019**, *42*, 969–981. [CrossRef]
32. Al-Refai, M.; Ali Hajji, M. Monotone iterative sequences for nonlinear boundary value problems of fractional order. *Nonlinear Anal.* **2011**, *74*, 3531–3539. [CrossRef]
33. Chen, C.; Bohner, M.; Jia, B. Method of upper and lower solutions for nonlinear Caputo fractional difference equations and its applications. *Fract. Calc. Appl. Anal.* **2019**, *22*, 1307–1320. [CrossRef]
34. Dhaigude, D.; Rizqan, B. Existence and uniqueness of solutions of fractional differential equations with deviating arguments under integral boundary conditions. *Kyungpook Math. J.* **2019**, *59*, 191–202.
35. Fazli, H.; Sun, H.; Aghchi, S. Existence of extremal solutions of fractional Langevin equation involving nonlinear boundary conditions. *Int. J. Comput. Math.* **2020**. [CrossRef]
36. Lin, X.; Zhao, Z. Iterative technique for a third-order differential equation with three-point nonlinear boundary value conditions. *Electron. J. Qual. Theory Differ. Equ.* **2016**, *12*, 10. [CrossRef]

37. Ma, K.; Han, Z.; Sun, S. Existence and uniqueness of solutions for fractional q-difference Schrödinger equations. *J. Appl. Math. Comput.* **2020**, *62*, 611–620. [CrossRef]
38. Mao, J.; Zhao, Z.; Wang, C. The unique iterative positive solution of fractional boundary value problem with q-difference. *Appl. Math. Lett.* **2020**, *100*, 106002. [CrossRef]
39. Meng, S.; Cui, Y. The extremal solution to conformable fractional differential equations involving integral boundary condition. *Mathematics* **2019**, *7*, 186. [CrossRef]
40. Wang, G.; Sudsutad, W.; Zhang, L.; Tariboon, J. Monotone iterative technique for a nonlinear fractional q-difference equation of Caputo type. *Adv. Diff. Equ.* **2016**, *2016*, 211. [CrossRef]
41. Yang, W. Monotone iterative technique for a coupled system of nonlinear Hadamard fractional differential equations. *J. Appl. Math. Comput.* **2019**, *59*, 585–596. [CrossRef]
42. Zhang, S. Monotone iterative method for initial value problem involving Riemann-Liouville fractional derivatives. *Nonlinear Anal.* **2009**, *71*, 2087–2093. [CrossRef]
43. Gorenflo, R.; Kilbas, A.A.; Mainardi, F.; Rogosin, S.V. *Mittag–Leffler Functions, Related Topics and Applications*; Springer: New York, NY, USA, 2014.
44. Diethelm, K.; Ford, N.J. Analysis of fractional differential equations. *J. Math. Anal. Appl.* **2002**, *265*, 229–248. [CrossRef]
45. Nieto, J.J. Maximum principles for fractional differential equations derived from Mittag-Leffler functions. *Appl. Math. Lett.* **2010**, *23*, 1248–1251. [CrossRef]
46. Royden, H.L. *Real Analysis*, 3rd ed.; Macmillan Publishing Company: New York, NY, USA, 1988.

© 2020 by the authors. Licensee MDPI, Basel, Switzerland. This article is an open access article distributed under the terms and conditions of the Creative Commons Attribution (CC BY) license (http://creativecommons.org/licenses/by/4.0/).

Article

Boundary Value Problem for Weak Nonlinear Partial Differential Equations of Mixed Type with Fractional Hilfer Operator

Tursun K. Yuldashev [1,*] and Bakhtiyor J. Kadirkulov [2]

[1] Uzbek-Israel Joint Faculty of High Technology and Engineering Mathematics, National University of Uzbekistan, Tashkent 100174, Uzbekistan
[2] Tashkent State Institute of Oriental Studies, Tashkent 100060, Uzbekistan; kadirkulovbj@gmail.com
* Correspondence: t.yuldashev@nuu.uz or tursun.k.yuldashev@gmail.com; Tel.: +998-99-519-59-31

Received: 20 May 2020; Accepted: 15 June 2020; Published: 17 June 2020

Abstract: In this paper, we consider a boundary value problem for a nonlinear partial differential equation of mixed type with Hilfer operator of fractional integro-differentiation in a positive rectangular domain and with spectral parameter in a negative rectangular domain. With respect to the first variable, this equation is a nonlinear fractional differential equation in the positive part of the considering segment and is a second-order nonlinear differential equation with spectral parameter in the negative part of this segment. Using the Fourier series method, the solutions of nonlinear boundary value problems are constructed in the form of a Fourier series. Theorems on the existence and uniqueness of the classical solution of the problem are proved for regular values of the spectral parameter. For irregular values of the spectral parameter, an infinite number of solutions of the mixed equation in the form of a Fourier series are constructed.

Keywords: mixed type nonlinear equation; boundary value problem; hilfer operator; mittag–leffler function; spectral parameter; solvability

1. Introduction

One of the most striking areas of mathematical analysis is the invention of fractional-order integro-differential operators. Today, the theory and application of operators of fractional differentiation and integration have become a powerful industry of theoretical and applied research at the highest levels of different science and technology. In particular, a concrete physical and engineering interpretation of the generalized fractional operator is given in [1] (Volume 4–8), [2–6]. At present, the operators of fractional differentiation and integration are also widely used in the study of problems associated with the study of the coronavirus COVID-19 (see, for example [1,7]).

In this paper we use Hilfer operator:

$$D^{\alpha,\gamma} = J_{0+}^{\gamma-\alpha} \frac{d}{dt} J_{0+}^{1-\gamma}, \quad 0 < \alpha \leq \gamma \leq 1,$$

where

$$J_{0+}^{\alpha} \varphi(t) = \frac{1}{\Gamma(\alpha)} \int_0^t \frac{\varphi(\tau)\,d\tau}{(t-\tau)^{1-\alpha}}, \quad \alpha > 0$$

is a Riemann–Liouville integral operator. For $\gamma = \alpha$ and $\gamma = 1$ we have $D^{\alpha,0} = {}_{RL}D_{0+}^{\alpha}$ and $D^{\alpha,1} = {}_{C}D_{0+}^{\alpha}$. Therefore, the generalized integro-differentiation operator $D^{\alpha,\gamma}$ is a continuous interpolation of the well-known fractional order differentiation operators of Riemann–Liouville and Caputo, which describe diffusion processes [1] (Volume 1, pp. 47–85).

Now we consider in detail a review of some works. For the first time, the generalized Riemann–Liouville operator (named as the Hilfer fractional derivative) was introduced by R. Hilfer on the basis of fractional time evolutions that arise during the transition from the microscopic scale to the macroscopic time scale [8]. Furthermore, R. Hilfer solved a Cauchy type problem for a fractional order equation with the same operator, applying in this case the Laplace transforms. In addition, using the integral Fourier, Laplace, and Mellin transforms, he investigated the Cauchy problem for the generalized diffusion equation, the solution of which is presented in the form of the Fox H-function.

It is applied in [9,10], the generalized fractional integro-differentiation operator in studying the dielectric relaxation in glass-forming liquids with different chemical compositions. For this, as usual, a classical Debye-type model was used, which describes exponential relaxation. The Debye-type model is determined by a first-order differential equation (see Equation (19) in [9]). But, as follows from the experiments, the ubiquitous feature of the dynamics of supercooled liquids and amorphous polymers is just non-exponential relaxation, which is the result of slow relaxation. To successfully describe the relaxation dynamics of glassy materials, the author of this article proposed a new model of dielectric relaxation containing derivatives and integrals of the non-integer order, which are a natural generalization of the Debye equation.

In [11] boundary value problems for the fractional diffusion equation with the time-generalized Riemann–Liouville fractional derivative (named as the Hilfer fractional derivative) in finite and infinite domains are studied. In the finite domain, the method of separation of variables and the Laplace transform method for solving the problem were used. In addition, the solution of the considered problem was obtained in the form of an infinite series containing the Mittag–Leffler function, and the asymptotic behavior of this solution at infinity was also found. In the infinite domain with respect to the spatial variable by the Fourier-Laplace transform method, the Cauchy problem is solved. In particular, a fundamental solution of the Cauchy problem is found and the fractional moments of the fundamental solution of the fractional diffusion equation are calculated. It is also shown in [11] that the corresponding solutions of the diffusion equations with fractional derivatives in the sense of Caputo or Riemann–Liouville are particular cases of diffusion equations with a fractional derivative according to Hilfer. The results obtained in this work are relevant in the study of the dielectric relaxation of glass and problems of the aquifer.

In [12] the analytical and numerical solution of boundary value problems for the fractional diffusion equation with the Hilfer fractional derivative was studied with respect to time and with respect to the Riesz–Feller spatial fractional derivative. To solve the problem, the Laplace and Fourier transform methods were used, and the solutions are presented by the Mittag–Leffler functions and the Fox H-function. A numerical solution of the problem is also considered by aid of approximating fractional derivatives with fractional derivatives of the Grunwald–Letnikov.

In [13], a new definition of the fractional derivative is introduced: The Hilfer–Prabhakar fractional derivative, which generalizes the fractional derivatives of Riemann–Liouville and Caputo. The new operator is constructed by replacing the Riemann–Liouville integrals of fractional order with more general Prabhakar integrals of fractional order. In addition, some applications of these generalized fractional derivatives in solving classical equations of mathematical physics are shown. Here we can note the heat equations and differential-difference equations that determine the dynamics of generalized random recovery processes, etc.

In [14] the properties of the Hilfer operator were investigated in a special functional space, and an operational method was developed for solving fractional differential equations with this operator. Developing the results of [14], the authors of [15] developed an operational method for solving fractional differential equations containing a finite linear combination of Hilfer operators with various parameters.

More detailed information as well as a bibliography related to the Hilfer fractional derivative can be found in the recently published monograph [16], where the theory of fractional integro-differentiation, including the Hilfer fractional derivative, is systematically presented. Section 2 of this paper gives the basic properties of the Hilfer operators, and its generalization is the Hilfer–Prabhakar fractional derivative, and Section 4 shows the applications of these fractional derivatives in solving various applied problems of mathematical physics.

So, a large number of scientific papers have been devoted to the investigation of initial, boundary, and inverse value problems for linear and nonlinear ordinary and partial differential equations (see also [17–26]).

We note that in [27] the problem of source identification was studied for the generalized diffusion equation with operator $D^{\alpha,\gamma}$. In the work [28] the inverse problems are investigated for a generalized fourth-order parabolic equation with the operator $D^{\alpha,\gamma}$.

In nature and in physics, processes that occur over time are usually nonlinear. Therefore, the study of nonlinear differential and functional-differential equations of fractional order is relevant.

2. Problem Statement

In a domain $\Omega = \{-a < t < b, \ 0 < x, y < l\}$ we consider a nonlinear partial fractional differential equation of mixed type:

$$0 = \begin{cases} \left(D^{\alpha,\gamma} - D^{\alpha,\gamma} \left(\frac{\partial^2}{\partial x^2} + \frac{\partial^2}{\partial y^2} \right) - \left(\frac{\partial^2}{\partial x^2} + \frac{\partial^2}{\partial y^2} \right) \right) U(t,x,y) \\ \quad - g_1(t) f_1 \left(x, y, \int_0^b \int_0^l \int_0^l \Theta_1(\theta,\zeta,\varsigma, \theta^{1-\gamma} U(\theta,\zeta,\varsigma)) \, d\theta \, d\zeta \, d\varsigma \right), \ (t,x,y) \in \Omega_1, \\ \left(\frac{\partial^2}{\partial t^2} - \frac{\partial^2}{\partial t^2} \left(\frac{\partial^2}{\partial x^2} + \frac{\partial^2}{\partial y^2} \right) - \omega^2 \left(\frac{\partial^2}{\partial x^2} + \frac{\partial^2}{\partial y^2} \right) \right) U(t,x,y) \\ \quad - g_2(t) f_2 \left(x, y, \int_{-a}^0 \int_0^l \int_0^l \Theta_2(\theta,\zeta,\varsigma, U(\theta,\zeta,\varsigma)) \, d\theta \, d\zeta \, d\varsigma \right), \ (t,x,y) \in \Omega_2, \end{cases} \quad (1)$$

where $\Omega_1 = \{0 < t < b, \ 0 < x, y < l\}$, $\Omega_2 = \{-a < t < 0, \ 0 < x, y < l\}$, ω is positive spectral parameter, and a, b are positive real numbers,

$$D^{\alpha,\gamma} = J_{0+}^{\gamma-\alpha} \frac{d}{dt} J_{0+}^{1-\gamma}, \ 0 < \alpha \leq \gamma \leq 1$$

is Hilfer operator, $g_1(t) \in C[0;b]$, $g_2(t) \in C[-a;0]$,

$$f_i(x,y,u) \in C([0;l]^2 \times \mathbb{R}), i = 1, 2, \ \Theta_1(t,x,y,U) \in C([0;b] \times [0;l]^2 \times \mathbb{R}),$$

$$\Theta_2(t,x,y,U) \in C([-a;0] \times [0;l]^2 \times \mathbb{R}).$$

Problem 1 (T_ω). *It is required to find a function $U(t,x,y)$, which belongs to the class:*

$$\begin{bmatrix} t^{1-\gamma} \frac{\partial^k U}{\partial x^k} \in C(\overline{\Omega}_1), \ t^{1-\gamma} \frac{\partial^k U}{\partial y^k} \in C(\overline{\Omega}_1), \ \frac{\partial^k U}{\partial x^k} \in C(\overline{\Omega}_2), \ \frac{\partial^k U}{\partial y^k} \in C(\overline{\Omega}_2), \\ D^{\alpha,\gamma} U \in C(\Omega_1), \ U_{tt}, \ U_{xx}, \ U_{yy} \in C(\Omega_1 \cup \Omega_2), \ k = 0, 1, 2, \end{bmatrix} \quad (2)$$

satisfies mixed differential Equation (1) in the domain $\Omega_1 \cup \Omega_2$, boundary value conditions:

$$U(t, 0, y) = U(t, l, y) = U(t, x, 0) = U(t, x, l) = 0, \ t \neq 0, \quad (3)$$

$$U(-a, x, y) = U(b, x, y) + \varphi(x, y), \ 0 \leq x, y \leq l, \quad (4)$$

gluing conditions:

$$\lim_{t\to+0} J_{0+}^{1-\gamma} U(t, x, y) = \lim_{t\to-0} U(t, x, y), \quad \lim_{t\to+0} J_{0+}^{1-\alpha} \frac{d}{dt} J_{0+}^{1-\gamma} U(t, x, y) = \lim_{t\to-0} \frac{d}{dt} U(t, x, y), \quad (5)$$

where $\varphi(x, y)$ is given a sufficiently smooth function.

Note that boundary value conditions of type (3) take place in modeling problems of the flow around a profile by a subsonic velocity stream with a supersonic zone. Nonlocal boundary value problems for different type of equations were studied in the works of many authors, in particular, in [29–36]. Nonlinear differential and integro-differential equations without mixing of the type of equations were studied in [37–42] by the Fourier series method.

In our work, unlike mixed parabolic-hyperbolic equations, the problem of small denominators do not arise. In this paper, we consider a boundary value problem for a mixed type nonlinear differential equation with Hilfer operator of fractional integro-differentiation. The Fourier method of separation of variables is used taking into account the features of the fractional integro-differentiation operator and nonlinearity. We study the solvability of problem (1)–(5) for various values of the spectral parameter. This work is a further development of the results of [35,38–40,42–45].

3. Nonhomogeneous Ordinary Differential Equation With Hilfer Operator

We consider the Cauchy problem for a nonhomogeneous differential equation of fractional order:

$$\begin{cases} D^{\alpha,\gamma} u(t) = k u(t) + f(t), \quad t \in (0, t_1), \\ \lim_{t\to+0} J_{0+}^{1-\gamma} u(t) = u_0, \end{cases} \quad (6)$$

where $f(t)$ is given continuous function, $u_0 = $ const.

Note that in [28], the Laplace method was applied to solve this problem. In [15], a solution was found using operational calculus for a more general problem than (6) in a specially constructed functional space. In our work, we use a more rational way to solve problem (6), which allows us to obtain an explicit solution.

We prove that there holds the following lemma.

Lemma 1. Let be $f(t) \in C(0; t_1] \cap L_1(0; t_1)$. Then the solution of the problem (6) $u(t) \in C(0; t_1] \cap L_1(0; t_1)$ is represented as follows:

$$u(t) = u_0 t^{\gamma-1} E_{\alpha,\gamma}(k t^\alpha) + \int_0^t (t-\tau)^{\alpha-1} E_{\alpha,\alpha}(k(t-\tau)^\alpha) f(\tau) d\tau, \quad (7)$$

where

$$E_{\alpha,\gamma}(z) = \sum_{m=0}^\infty \frac{z^m}{\Gamma(\alpha m + \gamma)}, \quad z, \alpha, \gamma \in \mathbb{C}, \ \operatorname{Re}(\alpha) > 0$$

is a Mittag–Leffler function (Volume 1, pp. 269–295) in [1].

Proof. We rewrite the differential equation of problem (6) in the form:

$$J_{0+}^{\gamma-\alpha} D_{0+}^{\gamma} u(t) = k u(t) + f(t).$$

Applying the operator J_{0+}^α to both sides of this equation, taking into account the linearity of this operator and the following formula [15]:

$$J_{0+}^\gamma D_{0+}^\gamma u(t) = u(t) - \frac{1}{\Gamma(\gamma)} J_{0+}^{1-\gamma} u(t)|_{t=0} t^{\gamma-1},$$

we obtain:
$$u(t) = \frac{u_0}{\Gamma(\gamma)} t^{\gamma-1} + J_{0+}^{\alpha} f(t) + k J_{0+}^{\alpha} u(t). \tag{8}$$

Using the lemma from [44], we represent the solution of Equation (8) as follows:
$$u(t) = \frac{u_0}{\Gamma(\gamma)} t^{\gamma-1} + J_{0+}^{\alpha} f(t) +$$
$$+ k \int_0^t (t-\tau)^{\alpha-1} E_{\alpha,\alpha}(k(t-\tau)^{\alpha}) \left[\frac{u_0}{\Gamma(\gamma)} \tau^{\gamma-1} + J_{0+}^{\alpha} f(\tau) \right] d\tau. \tag{9}$$

We rewrite the representation (9) as the sum of two expressions:
$$I_1(t) = u_0 \left[\frac{t^{\gamma-1}}{\Gamma(\gamma)} + \frac{k}{\Gamma(\gamma)} \int_0^t (t-\tau)^{\alpha-1} E_{\alpha,\alpha}(k(t-\tau)^{\alpha}) \tau^{\gamma-1} d\tau \right], \tag{10}$$

$$I_2(t) = J_{0+}^{\alpha} f(t) + k \int_0^t (t-\tau)^{\alpha-1} E_{\alpha,\alpha}(k(t-\tau)^{\alpha}) J_{0+}^{\alpha} f(\tau) d\tau. \tag{11}$$

We apply the following representations (Volume 1, pp. 269–295) in [1]:
$$E_{\alpha,\gamma}(z) = \frac{1}{\Gamma(\gamma)} + z E_{\alpha,\gamma+\alpha}(z), \quad \alpha > 0, \ \gamma > 0, \tag{12}$$

$$\frac{1}{\Gamma(\tau)} \int_0^z (z-t)^{\tau-1} E_{\alpha,\gamma}(kt^{\alpha}) t^{\gamma-1} dt = z^{\gamma+\tau-1} E_{\alpha,\gamma+\tau}(kz^{\alpha}), \quad \tau > 0, \ \gamma > 0. \tag{13}$$

Then for the integral (10) we obtain:
$$I_1(t) = u_0 t^{\gamma-1} E_{\alpha,\gamma}(kt^{\alpha}). \tag{14}$$

The integral in (11) we can transform as follows:
$$\int_0^t (t-\xi)^{\alpha-1} E_{\alpha,\alpha}(k(t-\xi)^{\alpha}) J_{0+}^{\alpha} f(\xi) d\xi$$
$$= \frac{1}{\Gamma(\alpha)} \int_0^t (t-\xi)^{\alpha-1} E_{\alpha,\alpha}(k(t-\xi)^{\alpha}) d\xi \int_0^{\tau} (\xi-s)^{\alpha-1} f(s) ds$$
$$= \frac{1}{\Gamma(\alpha)} \int_0^t f(s) ds \int_s^t (t-\xi)^{\alpha-1} (\xi-s)^{\alpha-1} E_{\alpha,\alpha}(k(t-\xi)^{\alpha}) d\xi. \tag{15}$$

Taking (13) into account the second integral in the last equality of (15) can be written as:
$$\int_s^t (t-\xi)^{\alpha-1} (\xi-s)^{\alpha-1} E_{\alpha,\alpha}(k(t-\xi)^{\alpha}) d\xi = \Gamma(\alpha)(t-\xi)^{2\alpha-1} E_{\alpha,2\alpha}(k(t-\xi)^{\alpha}).$$

Then, taking into account (12), we represent (11) in the following form:

$$I_2(t) = \int_0^t (t-\xi)^{\alpha-1} E_{\alpha,\alpha}\left(k(t-\xi)^\alpha\right) f(\xi)\, d\xi. \tag{16}$$

Substituting (14) and (16) into the sum $u(t) = I_1(t) + I_2(t)$, we obtain (7). The Lemma 1 is proved. □

4. Formal Expansion of the Solution of the Problem (1)–(5) into Fourier Series

The solution of the mixed differential Equation (1) in the domain Ω is sought in the form of a Fourier series:

$$U(t, x, y) = \sum_{n,m=1}^{\infty} u_{n,m}(t)\, \vartheta_{n,m}(x, y), \tag{17}$$

where

$$u_{n,m}(t) = \int_0^l \int_0^l U(t, x, y)\, \vartheta_{n,m}(x, y)\, dx\, dy, \tag{18}$$

$$\vartheta_{n,m}(x, y) = \frac{2}{l} \sin(\mu_n x) \sin(\mu_m x), \quad \mu_n = \frac{n\pi}{l}, \quad \mu_m = \frac{m\pi}{l}, \quad n, m \in \mathbb{N}.$$

We suppose also that:

$$f_i(x, y, \cdot) = \sum_{n,m=1}^{\infty} f_{in,m}(\cdot)\, \vartheta_{n,m}(x, y), \quad i = 1, 2, \tag{19}$$

where

$$f_{in,m}(\cdot) = \int_0^l \int_0^l f_i(x, y, \cdot)\, \vartheta_{n,m}(x, y)\, dx\, dy, \quad i = 1, 2.$$

Substituting series (17) and (19) into mixed Equation (1), we obtain a countable system of differential equations:

$$D^{\alpha,\gamma} u_{n,m}(t) + \lambda_{n,m}^2\, u_{n,m}(t) = g_1(t) f_{1n,m}(\cdot), \quad t > 0, \tag{20}$$

$$u''_{n,m}(t) + \lambda_{n,m}^2\, \omega^2 u_{n,m}(t) = g_2(t) f_{2n,m}(\cdot), \quad t < 0, \tag{21}$$

where

$$\lambda_{n,m}^2 = \frac{\mu_n^2 + \mu_m^2}{1 + \mu_n^2 + \mu_m^2}, \quad \mu_n = \frac{n\pi}{l}, \quad \mu_m = \frac{m\pi}{l}, \quad n, m \in \mathbb{N}.$$

Taking (18) into account from the conditions (5) we derive:

$$\lim_{t \to +0} J_{0+}^{1-\gamma} u_{n,m}(t) = \frac{2}{l} \int_0^l \int_0^l \lim_{t \to +0} J_{0+}^{1-\gamma} U(t, x, y) \sin(\mu_n x) \cdot \sin(\mu_m y)\, dx\, dy$$

$$= \frac{2}{l} \int_0^l \int_0^l \lim_{t \to -0} U(t, x, y) \sin(\mu_n x) \sin(\mu_m y)\, dx\, dy = \lim_{t \to -0} u_{n,m}(t), \tag{22}$$

$$\lim_{t \to +0} J_{0+}^{1-\alpha} \frac{d}{dt} J_{0+}^{1-\gamma} u_{n,m}(t) = \frac{2}{l} \int_0^l \int_0^l \lim_{t \to +0} J_{0+}^{1-\alpha} \frac{d}{dt} J_{0+}^{1-\gamma} U(t, x, y) \sin(\mu_n x) \sin(\mu_m y)\, dx\, dy$$

$$= \frac{2}{l} \int_0^l \int_0^l \lim_{t \to -0} \frac{d}{dt} U(t, x, y) \sin(\mu_n x) \sin(\mu_m y) \, dx \, dy = \lim_{t \to -0} \frac{d}{dt} u_{n,m}(t). \tag{23}$$

Analogously we find from condition (4) that:

$$u_{n,m}(-a) = u_{n,m}(b) + \varphi_{n,m}, \tag{24}$$

where

$$\varphi_{n,m} = \frac{2}{l} \int_0^l \int_0^l \varphi(x, y) \sin(\mu_n x) \sin(\mu_m y) \, dx \, dy, \quad n, m = 1, 2, \ldots$$

By applying Lemma 1, for (20) and (21) we obtain the general forms of solutions:

$$u_{n,m}(t) = A_{1n,m} t^{\gamma-1} E_{\alpha,\gamma}\left(-\lambda_{n,m}^2 t^\alpha\right) + f_{1n,m}(\cdot) h_{1n}(t), \quad t > 0, \tag{25}$$

$$u_{n,m}(t) = A_{2n,m} \sin \lambda_{n,m} \omega t + A_{3n,m} \cos \lambda_{n,m} \omega t + f_{2n,m}(\cdot) h_{2n,m}(t), \quad t < 0, \tag{26}$$

where $A_{in,m}$ are arbitrary constants, $i = \overline{1,3}$, $n, m = 1, 2, \ldots$

$$h_{1n,m}(t) = \int_0^t (t-s)^{\alpha-1} E_{\alpha,\alpha}\left(-\lambda_{n,m}^2 (t-s)^\alpha\right) g_1(s) \, ds,$$

$$h_{2n,m}(t) = \frac{1}{\lambda_{n,m} \omega} \int_0^t \sin(\lambda_{n,m} \omega (t-s)) g_2(s) \, ds.$$

Taking into account that $h_{1n,m}(0) = h_{2n,m}(0) = 0$ and satisfying functions (25) and (26) to conditions (22) and (23), we obtain the following systems of algebraic equations:

$$A_{2n,m} = -\frac{\lambda_{n,m}}{\omega} A_{1n,m}, \quad A_{3n,m} = A_{1n,m}. \tag{27}$$

Applying the condition (24) and representation (27) to (25) and (26), we derive:

$$A_{1n,m} = \frac{\varphi_{n,m} + f_{1n,m}(\cdot) h_{1n,m}(b) - f_{2n,m}(\cdot) h_{2n,m}(-a)}{\Delta_{n,m}(\omega)}, \tag{28}$$

if there holds the condition:

$$\Delta_{n,m}(\omega) = \lambda_{n,m} \omega^{-1} \sin(\lambda_{n,m} \omega a) + \cos(\lambda_{n,m} \omega a) - b^{\gamma-1} E_{\alpha,\gamma}\left(-\lambda_{n,m}^2 b^\alpha\right) \neq 0. \tag{29}$$

Substituting (28) into (27), for (25) and (26) we obtain the system of countable systems of nonlinear integral equations (SCSNIE):

$$u_{n,m}(t, \omega) = \mathbb{I}_1(t; u_{n,m})$$

$$\equiv \varphi_{n,m} \eta_{1n,m}(t, \omega) + f_{1n,m}(\cdot) \eta_{2n,m}(t, \omega) + f_{2n,m}(\cdot) \eta_{3n,m}(t, \omega), \quad t > 0, \tag{30}$$

$$u_{n,m}(t, \omega) = \mathbb{I}_2(t; u_{n,m})$$

$$\equiv \varphi_{n,m} \xi_{1n,m}(t, \omega) + f_{1n,m}(\cdot) \xi_{2n,m}(t, \omega) + f_{2n,m}(\cdot) \xi_{3n,m}(t, \omega), \quad t < 0, \tag{31}$$

where

$$f_{1n,m}(\cdot) = \int_0^l \int_0^l f_1\left(x, y, \int_0^b \int_0^l \int_0^l \Theta_1\left(\theta, \zeta, \varsigma, \sum_{i,j=1}^\infty \theta^{1-\gamma} u_{i,j}(\theta)\, \vartheta_{i,j}(\zeta, \varsigma)\right) d\theta\, d\zeta\, d\varsigma\right) \vartheta_{n,m}(x, y)\, dx\, dy,$$

$$f_{2n,m}(\cdot) = \int_0^l \int_0^l f_2\left(x, y, \int_{-a}^0 \int_0^l \int_0^l \Theta_2\left(\theta, \zeta, \varsigma, \sum_{i,j=1}^\infty u_{i,j}(\theta)\, \vartheta_{i,j}(\zeta, \varsigma)\right) d\theta\, d\zeta\, d\varsigma\right) \vartheta_{n,m}(x, y)\, dx\, dy,$$

$$\eta_{1n,m}(t, \omega) = \frac{t^{\gamma-1}}{\Delta_{n,m}(\omega)} E_{\alpha,\gamma}\left(-\lambda_{n,m}^2\, t^\alpha\right), \quad \eta_{2n,m}(t, \omega) = h_{1n,m}(t) + h_{1n,m}(b)\, \eta_{1n,m}(t, \omega),$$

$$\eta_{3n,m}(t, \omega) = -h_{2n,m}(-a)\, \eta_{1n,m}(t, \omega), \quad \xi_{1n,m}(t, \omega) = \frac{1}{\Delta_{n,m}(\omega)} \left[\sin(\lambda_{n,m}\omega t) + \cos(\lambda_{n,m}\omega t)\right],$$

$$\xi_{2n,m}(t, \omega) = h_{1n,m}(b)\, \xi_{1n,m}(t, \omega), \quad \xi_{3n,m}(t, \omega) = h_{2n,m}(t) + h_{2n,m}(-a)\, \xi_{1n,m}(t, \omega).$$

5. Solvability of SCSNIE (30) and (31)

Now we consider the case, when condition (29) is violated. Let $\Delta_{k,s}(\omega) = 0$ be for all ω. Then the considering problem ($\varphi(x, y) \equiv 0$) has the nontrivial solution:

$$V_{k,s}(t, x, y) = v_{k,s}(t)\, \vartheta_{k,s}(x, y), \quad (t, x, y) \in \Omega, \tag{32}$$

where

$$v_{k,s}(t) = \begin{cases} t^{\gamma-1} E_{\alpha,\gamma}\left(-\lambda_{k,s}^2\, t^\alpha\right) + f_{1k,s}(\cdot)\, h_{1k,s}(t), & t > 0, \\ \sin \lambda_{k,s}\omega t + \cos \lambda_{k,s}\omega t + f_{2k,s}(\cdot)\, h_{2k,s}(t), & t < 0. \end{cases}$$

From $\Delta_{n,m}(\omega) = 0$ we come to the trigonometric equation:

$$\sqrt{1 + \frac{\lambda_{n,m}^2}{\omega^2}} \sin(\lambda_{n,m}\omega a + \rho_{n,m}) - b^{\gamma-1} E_{\alpha,\gamma}\left(-\lambda_{n,m}^2\, b^\alpha\right) = 0, \tag{33}$$

where $\rho_{n,m} = \arcsin\left(\frac{\omega}{\sqrt{\omega^2 + \lambda_{n,m}^2}}\right)$. From this we obtain that the quantity $\Delta_{n,m}(\omega)$ vanishes at the values:

$$\omega = \frac{1}{\lambda_{n,m} a} \left[(-1)^z \arcsin \frac{\omega b^{\gamma-1} E_{\alpha,\gamma}\left(-\lambda_{n,m}^2\, b^\alpha\right)}{\sqrt{\omega^2 + \lambda_{n,m}^2}} + \pi z - \rho_{n,m}\right], \quad z \in \mathbb{N}.$$

The set of positive solutions $\Im$ of trigonometric Equation (33) with respect to spectral parameter ω is called a set of irregular values of the spectral parameter ω. The set of the remaining values of the spectral parameter $\aleph = (0; \infty) \setminus \Im$ is called a set of regular values of the spectral parameter ω. For all regular values of the spectral parameter ω, the quantity $\Delta_{n,m}(\omega)$ is nonzero. So, for large n, m the values of $\Delta_{n,m}(\omega)$ can not become quite small and there the problem of "small denominators" does not arise. Therefore, for regular values of the spectral parameter ω, the quantity $\Delta_{n,m}(\omega)$ is separated from zero.

Indeed, from the relations:

$$\lambda_{n,m}^2 = \frac{\mu_n^2 + \mu_m^2}{1 + \mu_n^2 + \mu_m^2}, \quad \mu_n = \frac{n\pi}{l}, \quad \mu_m = \frac{m\pi}{l}, \quad n, m \in \mathbb{N}$$

we see that $\lambda_{n,m}^2 \to 1$ as $n, m \to \infty$. Therefore, for regular values of the spectral parameter ω we have:

$$\lim_{n,m\to\infty} \Delta_{n,m}(\omega) = \frac{1}{\omega}\sin\omega a + \cos\omega a - b^{\gamma-1}E_{\alpha,\gamma}(-b^\alpha) \neq 0.$$

Lemma 2. *Suppose that $\gamma \in (0,1]$, a, b are arbitrary positive real numbers. Then for regular values of the spectral parameter $\omega \in \aleph$ and for arbitrary n, m there exists a positive constant M_0 such that there holds the following estimate:*

$$|\Delta_{n,m}(\omega)| \geq M_0 > 0. \tag{34}$$

Proof. From (33) for all n, m and a, $b > 0$ we derive:

$$|\Delta_{n,m}(\omega,\nu)| \geq \left|\pm\sqrt{1 + \frac{\lambda_{n,m}^2(\nu)}{\omega^2}} - b^{\gamma-1}E_{\alpha,\gamma}\left(-\lambda_{n,m}^2 b^\alpha\right)\right|$$

$$\geq \left|1 - b^{\gamma-1}E_{\alpha,\gamma}\left(-\lambda_{n,m}^2 b^\alpha\right)\right|.$$

We use the following properties of the Mittag–Leffler function (Volume 1, pp. 269–295) in [1]:

(1) For all $k > 0$, $\alpha, \gamma \in (0;1]$, $\alpha \leq \gamma$, $t \geq 0$ the function $t^{\gamma-1}E_{\alpha,\gamma}(-kt^\alpha)$ is completely monotonous and there holds:

$$(-1)^s \left[t^{\gamma-1}E_{\alpha,\gamma}(-kt^\alpha)\right]^{(s)} \geq 0, \quad s = 0,1,2,\ldots \tag{35}$$

(2) For all $\alpha \in (0;2)$, $\gamma \in \mathbb{R}$ and $\arg z = \pi$ there takes place the following estimate:

$$|E_{\alpha,\gamma}(z)| \leq \frac{M_1}{1+|z|}, \tag{36}$$

where $0 < M_1 = $ const does not depend from z.

Then, from the inequalities (35) and (36) we derive that there exists a number M_0 such that:

$$\left|1 - b^{\gamma-1}E_{\alpha,\gamma}\left(-\lambda_{n,m}^2 b^\alpha\right)\right| = M_0 > 0.$$

Consequently, for regular values of the spectral parameter ω there takes place (34): $|\Delta_{n,m}(\omega)| \geq M_0 > 0$. Lemma 2 is proved. □

Condition A. Let the following be fulfilled:

$$\varphi(x,y) \in C^3[0;l]^2, \varphi_{xxxx}(x,y) \in L_2[0;l]^2, \varphi_{yyyy}(x,y) \in L_2[0;l]^2.$$

Then by integrating in parts four times over the variable x the integral:

$$\varphi_{n,m} = \int_0^l \int_0^l \varphi(x,y)\vartheta_{n,m}(x,y)\,dx\,dy,$$

we derive that:

$$\varphi_{n,m} = \left(\frac{l}{\pi}\right)^4 \frac{\varphi_{n,m}^{(IV)}}{n^4}, \tag{37}$$

where,

$$\varphi_{n,m}^{(IV)} = \int_0^l \int_0^l \varphi_{xxxx}(x,y) \vartheta_{n,m}(x,y)\, dx\, dy, \qquad (38)$$

$$\vartheta_{n,m}(x,y) = \frac{2}{l} \sin \frac{\pi n}{l} x \sin \frac{\pi m}{l} y.$$

Similarly, by integrating the integral (38) in parts four times with respect to the variable y yields:

$$\varphi_{n,m}^{(IV)} = \left(\frac{l}{\pi}\right)^4 \frac{\varphi_{n,m}^{(VIII)}}{m^4}, \qquad (39)$$

where

$$\varphi_{n,m}^{(VIII)} = \int_0^l \int_0^l \varphi_{xxxxyyyy}(x,y) \vartheta_{n,m}(x,y)\, dx\, dy. \qquad (40)$$

Substituting (39) into (37), we obtain:

$$\varphi_{n,m} = \left(\frac{l}{\pi}\right)^8 \frac{\varphi_{n,m}^{(VIII)}}{n^4 m^4}. \qquad (41)$$

Applying the Bessel inequality for the integral (40), we obtain the estimate:

$$\sum_{n,m=1}^{\infty} \left[\varphi_{n,m}^{(VIII)}\right]^2 = \sum_{n,m=1}^{\infty} \left[\int_0^l \int_0^l \varphi_{xxxxyyyy}(x,y) \vartheta_{n,m}(x,y)\, dx\, dy\right]^2$$

$$\leq \int_0^l \int_0^l \left[\varphi_{xxxxyyyy}(x,y)\right]^2 dx\, dy < \infty. \qquad (42)$$

Condition B. Let the following be fulfilled:

$$f_i(x,y,u) \in C_{x,y,u}^{3,3,0}\left([0;l]^2 \times \mathbb{R}\right), \quad f_{ixxxx}(x,y,u) \in L_2\left([0;l]^2 \times \mathbb{R}\right),$$

$$f_{iyyyy}(x,y,u) \in L_2\left([0;l]^2 \times \mathbb{R}\right), \quad i = 1, 2,$$

where

$$L_2\left([0;l]^2 \times \mathbb{R}\right) = \left\{f(x,y,u): \sqrt{\int_0^l \int_0^l |f(x,y,u)|^2 dx\, dy} < \infty\right\}.$$

Similarly to the case of condition **A**, we obtain:

$$f_{in,m}(\cdot) = \left(\frac{l}{\pi}\right)^8 \frac{f_{in,m}^{(VIII)}(\cdot)}{n^4 m^4}, \qquad (43)$$

$$\sum_{n,m=1}^{\infty} \left[f_{in,m}^{(VIII)}(\cdot)\right]^2 \leq \int_0^l \int_0^l \left[f_{ixxxxyyyy}(x,y,\cdot)\right]^2 dx\, dy < \infty, \qquad (44)$$

where

$$f_{in,m}^{(VIII)}(\cdot) = \frac{2}{l} \int_0^l \int_0^l f_{ixxxxyyyy}(x,y,\cdot) \sin \frac{\pi n}{l} x \sin \frac{\pi m}{l} y\, dx\, dy.$$

For all regular values of the spectral parameter $\omega \in \aleph$ the SCSNIE (30) and (31) are true. In order to prove the unique solvability of SCSNIE (30) and (31), we introduce the following well-knowing spaces.

Space $B_2[-a;b]$ of sequences of continuous functions $\{u_{n,m}(t)\}_{n,m=1}^{\infty}$ on the segment $[-a;b]$ with the norm:

$$\|u(t)\|_{B_2[-a;b]} = \|u(t)\|_{B_2[-a;0]} + \|u(t)\|_{B_2[0;b]}$$

$$= \sqrt{\sum_{n,m=1}^{\infty}\left(\max_{t\in[-a;0]}|u_{n,m}(t)|\right)^2} + \sqrt{\sum_{n,m=1}^{\infty}\left(\max_{t\in[0;b]}|u_{n,m}(t)|\right)^2} < \infty.$$

The space $L_2[0;l]^2$ of square-summable functions on the domain $[0;l]^2 = [0;l] \times [0;l]$ with the norm:

$$\|\vartheta(x,y)\|_{L_2[0;l]^2} = \sqrt{\int_0^l \int_0^l |\vartheta(x,y)|^2 dx\,dy} < \infty.$$

On the basis of lemma 2, Conditions **A** and **B** for regular spectral values from the sets $\aleph$ we prove that it holds.

Theorem 1. *Suppose that the following conditions and Conditions A, B are fulfilled:*

(1) $\chi_{11} = \max\limits_{i=\overline{1,3}}\max\limits_{n,m\in \mathbb{N}}\max\limits_{t\in[0;b]}|t^{1-\gamma}\eta_{inm}(t,\omega)| < \infty$; $\chi_{21} = \max\limits_{i=\overline{1,3}}\max\limits_{n,m\in \mathbb{N}}\max\limits_{t\in[-a;0]}|\xi_{inm}(t,\omega)| < \infty$;

(2) $\chi_{30} = \|\varphi_{xxxxyyyy}(x,y)\|_{L_2[0;l]^2} < \infty$; $\chi_{3i} = \|f_{ixxxxyyyy}(x,y,\gamma)\|_{L_2[0;l]^2} < \infty$;

(3) $|f_{ixxxxyyyy}(x,y,\gamma_1) - f_{ixxxxyyyy}(x,y,\gamma_2)| \leq K_i(x,y)|\gamma_1 - \gamma_2|$,

$K_{0i} = \|K_i(x,y)\|_{L_2[0;l]^2} < \infty$;

(4) $|\Theta_i(\xi,x,y,u_1) - \Theta_i(\xi,x,y,u_2)| \leq \Theta_{1i}(x,y)|u_1 - u_2|$,

$\Theta_{2i} = \|\Theta_{1i}(x,y)\|_{L_2[0;l]^2} < \infty$, $i = 1, 2$;

(5) $\rho = \gamma_2(\gamma_1 + \gamma_3)\gamma_4 < 1$, $\gamma_4 = \max\{bK_{01}\Theta_{21}; aK_{02}\Theta_{22}\}$.

Then SCSNIE (30) and (31) are uniquely solvable in the spaces $B_2[-a;0]$ and $B_2[0;b]$, respectively for all regular values of the spectral parameter $\omega \in \aleph$.

Proof. We use the method of compressing mappings in the Banach spaces $B_2[-a;0]$ and $B_2[0;b]$. Successive approximations are defined as follows:

$$\begin{cases} u_{n,m}^0(t,\omega) = \varphi_{n,m}\eta_{1n,m}(t,\omega),\ u_{n,m}^{k+1} = \mathbb{I}_1(t;u_{n,m}^k),\ k=0,1,2,\ldots,\ t > 0, \\ u_{n,m}^0(t,\omega) = \varphi_{n,m}\xi_{1n,m}(t,\omega),\ u_{n,m}^{k+1} = \mathbb{I}_2(t;u_{n,m}^k),\ t < 0,\ \omega \in \aleph. \end{cases} \quad (45)$$

When $t > 0$, by virtue of the first condition of the theorem and applying Cauchy–Schwarz inequality and properties (41) and (42) to the approximations (45) for the zero approximation $u_{n,m}^0(t,\omega)$ with the norm in $B_2[0;b]$ obtains the estimate:

$$\left\|t^{1-\gamma}u^0(t,\omega)\right\|_{B_2[0;b]} \leq \max_{t\in[0;b]}\sum_{n,m=1}^{\infty}|\varphi_n|\left|t^{1-\gamma}\eta_{1nm}(t,\omega)\right|$$

$$\leq \max_{n,m\in\mathbb{N}}\max_{t\in[0;b]}\left|t^{1-\gamma}\eta_{1nm}(t,\omega)\right|\sum_{n,m=1}^{\infty}\left|\left(\frac{l}{\pi}\right)^8 \frac{\varphi_{n,m}^{(VIII)}}{n^4 m^4}\right|$$

$$\leq \chi_{11}\left(\frac{l}{\pi}\right)^8 \sum_{n,m=1}^{\infty}\left|\frac{1}{n^4 m^4}\right|\cdot\left|\varphi_{n,m}^{(VIII)}\right| \leq \gamma_1 \sqrt{\sum_{n,m=1}^{\infty}\frac{1}{n^8 m^8}}\sqrt{\sum_{n,m=1}^{\infty}\left|\varphi_{n,m}^{(VIII)}\right|^2}$$

$$\leq \gamma_1 \gamma_2 \sqrt{\int_0^l \int_0^l [\varphi_{xxxxyyyy}(x,y)]^2 \, dx \, dy} = \gamma_1 \gamma_2 \gamma_{30} < \infty, \tag{46}$$

where $\gamma_1 = \chi_{11} \left(\frac{l}{\pi}\right)^8$, $\gamma_2 = \sqrt{\sum_{n,m=1}^{\infty} \frac{1}{n^8 m^8}}$.

Similarly, by virtue of the conditions of the theorem and applying Cauchy–Schwarz inequality and properties (43) and (44) for the first difference of approximations (45), we derive:

$$\left\| t^{1-\gamma} \left(u^1(t,\omega) - u^0(t,\omega)\right) \right\|_{B_2[0;b]} \leq \max_{t \in [0;b]} \sum_{n,m=1}^{\infty} \left| f_{1n,m}^0(\cdot) \right| \cdot \left| t^{1-\gamma} \eta_{2nm}(t,\omega) \right|$$

$$+ \max_{t \in [0;b]} \sum_{n,m=1}^{\infty} \left| f_{2n,m}^0(\cdot) \right| \cdot \left| t^{1-\gamma} \eta_{3nm}(t,\omega) \right|$$

$$\leq \max_{n,m \in \mathbb{N}} \max_{t \in [0;b]} \left| t^{1-\gamma} \eta_{2nm}(t,\omega) \right| \left(\frac{l}{\pi}\right)^8 \sum_{n,m=1}^{\infty} \frac{\left| f_{1n,m}^{(VIII)}(\cdot) \right|}{n^4 m^4}$$

$$+ \max_{n,m \in \mathbb{N}} \max_{t \in [0;b]} \left| t^{1-\gamma} \eta_{3nm}(t,\omega) \right| \left(\frac{l}{\pi}\right)^8 \sum_{n,m=1}^{\infty} \frac{\left| f_{2n,m}^{(VIII)}(\cdot) \right|}{n^4 m^4}$$

$$\leq \chi_{11} \left(\frac{l}{\pi}\right)^8 \left[\sum_{n,m=1}^{\infty} \frac{1}{n^4 m^4} \left| f_{1n,m}^{(VIII)}(\cdot) \right| + \sum_{n,m=1}^{\infty} \frac{1}{n^4 m^4} \left| f_{2n,m}^{(VIII)}(\cdot) \right| \right]$$

$$\leq \gamma_1 \sqrt{\sum_{n,m=1}^{\infty} \frac{1}{n^8 m^8}} \left[\sqrt{\sum_{n,m=1}^{\infty} \left| f_{1n,m}^{(VIII)}(\cdot) \right|^2} + \sqrt{\sum_{n,m=1}^{\infty} \left| f_{2n,m}^{(VIII)}(\cdot) \right|^2} \right]$$

$$\leq \gamma_1 \gamma_2 \left[\sqrt{\int_0^l \int_0^l [f_{1xxxxyyyy}(x,y,\cdot)]^2 \, dx \, dy} \right.$$

$$\left. + \sqrt{\int_0^l \int_0^l [f_{2xxxxyyyy}(x,y,\cdot)]^2 \, dx \, dy} \right] = \gamma_1 \gamma_2 (\chi_{31} + \chi_{32}) < \infty, \tag{47}$$

where

$$f_{1n,m}^k(\cdot) = \int_0^l \int_0^l f_1 \left(x, y, \int_0^b \int_0^l \int_0^l \Theta_1 \left(\theta, \zeta, \varsigma, \sum_{i,j=1}^{\infty} \theta^{1-\gamma} u_{i,j}^k(\theta) \vartheta_{i,j}(\zeta,\varsigma) \right) d\theta \, d\zeta \, d\varsigma \right) \vartheta_{n,m}(x,y) \, dx \, dy,$$

$$f_{2n,m}^k(\cdot) = \int_0^l \int_0^l f_2 \left(x, y, \int_{-a}^0 \int_0^l \int_0^l \Theta_2 \left(\theta, \zeta, \varsigma, \sum_{i,j=1}^{\infty} u_{i,j}^k(\theta) \vartheta_{i,j}(\zeta,\varsigma) \right) d\theta \, d\zeta \, d\varsigma \right) \vartheta_{n,m}(x,y) \, dx \, dy,$$

$k = 0, 1, 2, \ldots$

We use the conditions of theorem, Cauchy–Schwarz inequality, and Bessel inequality for the arbitrary difference $u_{n,m}^{k+1}(t,\omega) - u_{n,m}^k(t,\omega)$ with the norm in $B_2[0;b]$. Then we derive from (45) the following estimate:

$$\left\| t^{1-\gamma} \left(u^{k+1}(t,\omega) - u^k(t,\omega)\right) \right\|_{B_2[0;b]}$$

$$\leq \max_{t\in[0;b]} \sum_{n,m=1}^{\infty} \left| f_{1n,m}^{k}(\cdot) - f_{1n,m}^{k-1}(\cdot) \right| \left| t^{1-\gamma} \eta_{2nm}(t,\omega) \right|$$

$$+ \max_{t\in[0;b]} \sum_{n,m=1}^{\infty} \left| f_{2n,m}^{k}(\cdot) - f_{2n,m}^{k-1}(\cdot) \right| \cdot \left| t^{1-\gamma} \eta_{3nm}(t,\omega) \right|$$

$$\leq \gamma_1 \left[\sum_{n,m=1}^{\infty} \frac{1}{n^4 m^4} \left| f_{1n,m}^{k(VIII)}(\cdot) - f_{1n,m}^{k-1(VIII)}(\cdot) \right| + \sum_{n,m=1}^{\infty} \frac{1}{n^4 m^4} \left| f_{2n,m}^{k(VIII)}(\cdot) - f_{2n,m}^{k-1(VIII)}(\cdot) \right| \right]$$

$$\leq \gamma_1 \sqrt{ \sum_{n,m=1}^{\infty} \frac{1}{n^8 m^8} } \left[\sqrt{ \int_0^l \int_0^l \left| f_{1xxxxyyyy}^{k}(x,y,\cdot) - f_{1xxxxyyyy}^{k-1}(x,y,\cdot) \right|^2 dx\,dy } \right.$$

$$\left. + \sqrt{ \int_0^l \int_0^l \left| f_{2xxxxyyyy}^{k}(x,y,\cdot) - f_{2xxxxyyyy}^{k-1}(x,y,\cdot) \right|^2 dx\,dy } \right]$$

$$\leq \gamma_1 \gamma_2 \left[\sqrt{ \int_0^l \int_0^l |K_1(x,y)|^2 dx\,dy \int_0^b \int_0^l \int_0^l \left| \Theta_1^k(\cdot) - \Theta_1^{k-1}(\cdot) \right| d\theta\,d\zeta\,d\varsigma } \right.$$

$$\left. + \sqrt{ \int_0^l \int_0^l |K_2(x,y)|^2 dx\,dy \int_{-a}^0 \int_0^l \int_0^l \left| \Theta_2^k(\cdot) - \Theta_2^{k-1}(\cdot) \right| d\theta\,d\zeta\,d\varsigma } \right]$$

$$\leq \gamma_1 \gamma_2 \left[K_{01} \int_0^b \int_0^l \int_0^l |\Theta_{11}(\zeta,\varsigma)| \cdot \left| \sum_{i,j=1}^{\infty} \theta^{1-\gamma} \left[u_{i,j}^k(\theta) - u_{i,j}^{k-1}(\theta) \right] \vartheta_{i,j}(\zeta,\varsigma) \right| d\theta\,d\zeta\,d\varsigma \right.$$

$$\left. + K_{02} \int_{-a}^0 \int_0^l \int_0^l |\Theta_{12}(\zeta,\varsigma)| \cdot \left| \sum_{i,j=1}^{\infty} \left[u_{i,j}^k(\theta) - u_{i,j}^{k-1}(\theta) \right] \vartheta_{i,j}(\zeta,\varsigma) \right| d\theta\,d\zeta\,d\varsigma \right]$$

$$\leq \gamma_1 \gamma_2 \left[K_{01} \|\Theta_{11}(x,y)\|_{L_2[0;l]^2} \int_0^b \left\| \theta^{1-\gamma} \left(u^k(\theta,\omega) - u^{k-1}(\theta,\omega) \right) \right\|_{B_2[0;b]} d\theta \right.$$

$$\left. + K_{02} \|\Theta_{12}(x,y)\|_{L_2[0;l]^2} \int_{-a}^0 \left\| u^k(\theta,\omega) - u^{k-1}(\theta,\omega) \right\|_{B_2[-a;0]} d\theta \right]$$

$$\leq \gamma_1 \gamma_2 \left[b K_{01} \Theta_{21} \left\| t^{1-\gamma} \left(u^k(t,\omega) - u^{k-1}(t,\omega) \right) \right\|_{B_2[0;b]} \right.$$

$$\left. + a K_{02} \Theta_{22} \left\| u^k(t,\omega) - u^{k-1}(t,\omega) \right\|_{B_2[-a;0]} \right]. \tag{48}$$

When $t < 0$, by virtue of the conditions of the theorem and applying the Cauchy–Schwarz inequality and Bessel inequality to (45) we similarly obtain the following estimates:

$$\left\| u^0(t,\omega) \right\|_{B_2[-a;0]} \leq \max_{t\in[-a;0]} \sum_{n,m=1}^{\infty} |\varphi_n| |\xi_{1nm}(t,\omega)|$$

$$\leq \gamma_2 \gamma_3 \|\varphi_{xxxxyyyy}(x,y)\|_{L_2[0;l]^2} < \infty, \tag{49}$$

where $\gamma_3 = \chi_{21}\left(\frac{l}{\pi}\right)^8$;

$$\left\|u^1(t,\omega) - u^0(t,\omega)\right\|_{B_2[-a;0]} \leq \max_{t\in[-a;0]} \sum_{n,m=1}^{\infty} \left|f^0_{1n,m}(\cdot)\right| \cdot \left|\xi_{2nm}(t,\omega)\right|$$

$$+ \max_{t\in[-a;0]} \sum_{n,m=1}^{\infty} \left|f^0_{2n,m}(\cdot)\right| \cdot \left|\xi_{3nm}(t,\omega)\right| \leq \gamma_2 \gamma_3 \left[\left\|f_{1xxxxyyyy}(x,y,\cdot)\right\|_{L_2[0;l]^2}\right.$$

$$\left. + \left\|f_{2xxxxyyyy}(x,y,\cdot)\right\|_{L_2[0;l]^2}\right] = \gamma_2 \gamma_3 \left(\chi_{31} + \chi_{32}\right) < \infty; \qquad (50)$$

$$\left\|u^{k+1}(t,\omega) - u^k(t,\omega)\right\|_{B_2[-a;0]} \leq \max_{t\in[-a;0]} \sum_{n,m=1}^{\infty} \left|f^k_{1n,m}(\cdot) - f^{k-1}_{1n,m}(\cdot)\right| \left|\xi_{2nm}(t,\omega)\right|$$

$$+ \max_{t\in[-a;0]} \sum_{n,m=1}^{\infty} \left|f^k_{2n,m}(\cdot) - f^{k-1}_{2n,m}(\cdot)\right| \cdot \left|\xi_{3nm}(t,\omega)\right|$$

$$\leq \gamma_2 \gamma_3 \left[\left\|K_1(x,y)\right\|_{L_2[0;l]^2} \int_0^b \int_0^l \int_0^l \left|\Theta^k_1(\cdot) - \Theta^{k-1}_1(\cdot)\right| d\theta d\zeta d\varsigma\right.$$

$$\left. + \left\|K_2(x,y)\right\|_{L_2[0;l]^2} \int_{-a}^0 \int_0^l \int_0^l \left|\Theta^k_2(\cdot) - \Theta^{k-1}_2(\cdot)\right| d\theta d\zeta d\varsigma\right]$$

$$\leq \gamma_2 \gamma_3 \left[K_{01} \left\|\Theta_{11}(x,y)\right\|_{L_2[0;l]^2} \int_0^b \left\|\theta^{1-\gamma}\left(u^k(\theta,\omega) - u^{k-1}(\theta,\omega)\right)\right\|_{B_2[0;b]} d\theta\right.$$

$$\left. + K_{02} \left\|\Theta_{12}(x,y)\right\|_{L_2[0;l]^2} \int_{-a}^0 \left\|u^k(\theta,\omega) - u^{k-1}(\theta,\omega)\right\|_{B_2[-a;0]} d\theta\right]$$

$$\leq \gamma_2 \gamma_3 \left[b K_{01} \Theta_{21} \left\|t^{1-\gamma}\left(u^k(t,\omega) - u^{k-1}(t,\omega)\right)\right\|_{B_2[0;b]}\right.$$

$$\left. + a K_{02} \Theta_{22} \left\|u^k(t,\omega) - u^{k-1}(t,\omega)\right\|_{B_2[-a;0]}\right]. \qquad (51)$$

Adding inequalities (48) and (51), we obtain:

$$\left\|u^{k+1}(t,\omega) - u^k(t,\omega)\right\|_{B_2[-a;b]} \leq \rho \left\|u^k(t,\omega) - u^{k-1}(t,\omega)\right\|_{B_2[-a;b]}, \qquad (52)$$

where $\rho = \gamma_2 (\gamma_1 + \gamma_3) \gamma_4$, $\gamma_4 = \max\{b K_{01} \Theta_{21}; a K_{02} \Theta_{22}\}$.

According to the last condition of the theorem there is $\rho = \gamma_2 (\gamma_1 + \gamma_3) \gamma_4 < 1$. Therefore from the estimates (46), (47), (49), (50) and (52) implies that the operators on the right side of (30), and (31) are compressive and there exists a unique fixed point for these operators. Therefore the SCSNIE (30) and (31) are uniquely solvable in the space $B_2[-a;b]$ for regular spectral values of parameter $\omega \in \aleph$. Theorem 1 is thus proved. □

6. Convergence of Fourier Series

Substituting SCSNIE (30) and (31) into the Fourier series (17), we obtain:

$$U(t,x,y,\omega) = \sum_{n,m=1}^{\infty} \vartheta_{n,m}(x,y) \left[\varphi_{n,m} \eta_{1n,m}(t,\omega) + \eta_{2n,m}(t,\omega) f_{1n,m}(\cdot)\right.$$

$$+\eta_{3n,m}(t,\omega) f_{2n,m}(\cdot)], \quad (t,x,y) \in \Omega_1, \tag{53}$$

$$U(t,x,y,\omega) = \sum_{n,m=1}^{\infty} \vartheta_{n,m}(x,y) \left[\varphi_{n,m} \zeta_{1n,m}(t,\omega) + \zeta_{2n,m}(t,\omega) f_{1n,m}(\cdot)\right.$$

$$\left.+\zeta_{3n,m}(t,\omega) f_{2n,m}(\cdot)\right], \quad (t,x,y) \in \Omega_2, \tag{54}$$

where

$$f_{1n,m}(\cdot) = \int_0^l \int_0^l f_1\left(x,y, \int_0^b \int_0^l \int_0^l \Theta_1\left(\theta,\zeta,\varsigma, \sum_{i,j=1}^{\infty} \theta^{1-\gamma} u_{i,j}(\theta) \vartheta_{i,j}(\zeta,\varsigma)\right) d\theta d\zeta d\varsigma\right) \vartheta_{n,m}(x,y) dx dy,$$

$$f_{2n,m}(\cdot) = \int_0^l \int_0^l f_2\left(x,y, \int_{-a}^0 \int_0^l \int_0^l \Theta_2\left(\theta,\zeta,\varsigma, \sum_{i,j=1}^{\infty} u_{i,j}(\theta) \vartheta_{i,j}(\zeta,\varsigma)\right) d\theta d\zeta d\varsigma\right) \vartheta_{n,m}(x,y) dx dy.$$

Theorem 2. *Let conditions of the Theorem 1 be fulfilled. Then for regular values of the spectral parameter $\omega \in \aleph$ the Fourier series (53) and (54) are convergent absolutely and uniformly in the domain Ω_1 and Ω_2, respectively. The series (53) and (54) possess the Properties (2).*

Proof. We prove the absolutely and uniformly convergence of series (53) and (54). Similarly to the estimates (46), (47) and (49), (50), we obtain:

$$\left|t^{1-\gamma} U(t,x,y,\omega)\right| \leq \max_{t \in [0;b]} \sum_{n,m=1}^{\infty} \left|t^{1-\gamma} u_{n,m}(t,\omega)\right| \cdot |\vartheta_{nm}(x,y)|$$

$$\leq \frac{2}{l} \max_{t \in [0;b]} \left[\sum_{n,m=1}^{\infty} |\varphi_{n,m}(\cdot)| \cdot \left|t^{1-\gamma} \eta_{1nm}(t,\omega)\right| + \sum_{n,m=1}^{\infty} |f_{1n,m}(\cdot)| \cdot \left|t^{1-\gamma} \eta_{2nm}(t,\omega)\right|\right.$$

$$\left.+ \sum_{n,m=1}^{\infty} |f_{2n,m}(\cdot)| \cdot \left|t^{1-\gamma} \eta_{3nm}(t,\omega)\right|\right]$$

$$\leq \frac{2}{l} \gamma_1 \left[\sum_{n,m=1}^{\infty} \frac{1}{n^4 m^4} \left|\varphi_{n,m}^{(VIII)}\right| + \sum_{n,m=1}^{\infty} \frac{1}{n^4 m^4} \left|f_{1n,m}^{(VIII)}(\cdot)\right| + \sum_{n,m=1}^{\infty} \frac{1}{n^4 m^4} \left|f_{2n,m}^{(VIII)}(\cdot)\right|\right]$$

$$\leq \gamma_5 \left[\|\varphi_{xxxxyyyy}(x,y)\|_{L_2[0;l]^2} + \|f_{1xxxxyyyy}(x,y,\cdot)\|_{L_2[0;l]^2}\right.$$

$$\left.+ \|f_{2xxxxyyyy}(x,y,\cdot)\|_{L_2[0;l]^2}\right] = \gamma_5 (\chi_{30} + \chi_{31} + \chi_{32}) < \infty, \quad \gamma_5 = \frac{2}{l} \gamma_1 \gamma_2; \tag{55}$$

$$|U(t,x,y,\omega)| \leq \max_{t \in [-a;0]} \sum_{n,m=1}^{\infty} |u_{n,m}(t,\omega)| \cdot |\vartheta_{nm}(x,y)|$$

$$\leq \frac{2}{l} \max_{t \in [-a;0]} \left[\sum_{n,m=1}^{\infty} |\varphi_{n,m}(\cdot)| \cdot |\zeta_{1nm}(t,\omega)| + \sum_{n,m=1}^{\infty} |f_{1n,m}(\cdot)| \cdot |\zeta_{2nm}(t,\omega)|\right.$$

$$\left.+ \sum_{n,m=1}^{\infty} |f_{2n,m}(\cdot)| \cdot |\zeta_{3nm}(t,\omega)|\right] \leq \frac{2}{l} \gamma_3 \left[\sum_{n,m=1}^{\infty} \frac{1}{n^4 m^4} \left|\varphi_{n,m}^{(VIII)}\right|\right.$$

$$\left.+ \sum_{n,m=1}^{\infty} \frac{1}{n^4 m^4} \left|f_{1n,m}^{(VIII)}(\cdot)\right| + \sum_{n,m=1}^{\infty} \frac{1}{n^4 m^4} \left|f_{2n,m}^{(VIII)}(\cdot)\right|\right]$$

$$\leq \gamma_6 (\chi_{30} + \chi_{31} + \chi_{32}) < \infty, \quad \gamma_6 = \frac{2}{l} \gamma_2 \gamma_3. \tag{56}$$

Similarly to case of (55) and (56), it is easy to prove that the following series are convergent absolutely and uniformly in the domain Ω_1 and Ω_2, respectively:

$$t^{1-\gamma} D^{\alpha,\gamma} U(t,x,y,\omega) = \sum_{n,m=1}^{\infty} t^{1-\gamma} D^{\alpha,\gamma} u_{n,m}(t,\omega) \vartheta_{n,m}(x,y), \quad (t,x,y) \in \Omega_1, \tag{57}$$

$$t^{1-\gamma} \frac{\partial^k U(t,x,y,\omega)}{\partial x^k} = (-1)^{k+1} \sum_{n,m=1}^{\infty} t^{1-\gamma} u_{n,m}(t,\omega) \mu_n^k \vartheta_{n,m}(x,y), \quad k=1,2, \quad (t,x,y) \in \Omega_1, \tag{58}$$

$$t^{1-\gamma} \frac{\partial^k U(t,x,y,\omega)}{\partial y^k} = (-1)^{k+1} \sum_{n,m=1}^{\infty} t^{1-\gamma} u_{n,m}(t,\omega) \mu_m^k \vartheta_{n,m}(x,y), \quad k=1,2, \quad (t,x,y) \in \Omega_1, \tag{59}$$

$$\frac{\partial^2 U(t,x,y,\omega)}{\partial t^2} = \sum_{n,m=1}^{\infty} \frac{d^2 u_{n,m}(t,\omega)}{dt^2} \vartheta_{n,m}(x,y), \quad (t,x,y) \in \Omega_2, \tag{60}$$

$$\frac{\partial^k U(t,x,y,\omega)}{\partial x^k} = (-1)^{k+1} \sum_{n,m=1}^{\infty} u_{n,m}(t,\omega) \mu_n^k \vartheta_{n,m}(x,y), \quad k=1,2, \quad (t,x,y) \in \Omega_2, \tag{61}$$

$$\frac{\partial^k U(t,x,y,\omega)}{\partial y^k} = (-1)^{k+1} \sum_{n,m=1}^{\infty} u_{n,m}(t,\omega) \mu_m^k \vartheta_{n,m}(x,y), \quad k=1,2, \quad (t,x,y) \in \Omega_2. \tag{62}$$

Theorem 2 is proved. □

7. Irregular Value of Spectral Parameter ω

We note that $\Delta_{n,m}(\omega) = 0$ for irregular values of the spectral parameter $\omega \in \Im$ and $n, m = k, s$ ($\gamma \neq 1$). Then, for the solvability of systems (25) and (26), it is necessary and sufficient that the orthogonality conditions are satisfied:

$$\varphi_{k,s} = \int_0^l \int_0^l \varphi(x,y) \vartheta_{k,s}(x,y) \, dx \, dy = 0. \tag{63}$$

In this case, by virtue of (32), the solutions of nonlocal problem are represented as:

$$U(t,x,y) = \sum_{k,s=1}^{\infty} C_{k,s} \left[t^{\gamma-1} E_{\alpha,\gamma}\left(-\lambda_{k,s}^2 t^\alpha\right) + f_{1k,s}(\cdot) h_{1k,s}(t) \right] \vartheta_{k,s}(x,y), \quad (t,x,y) \in \Omega_1, \tag{64}$$

$$U(t,x,y) = \sum_{k,s=1}^{\infty} C_{k,s} \left[\sin \lambda_{k,s} \omega t + \cos \lambda_{k,s} \omega t + f_{2k,s}(\cdot) h_{2k,s}(t) \right] \vartheta_{k,s}(x,y), \quad (t,x,y) \in \Omega_2, \tag{65}$$

where $k, s = k_1, ..., k_s$, $C_{k,s}$ are arbitrary constants.

The absolute and uniform convergence of the obtained series (64) and (65) is clear, since $C_{k,s}$ are arbitrary numbers. Them we can select that these series converge. We recall that the Fourier coefficient functions $f_{1k,s}(\cdot)$ and $f_{2k,s}(\cdot)$ in (64) and (65) satisfy the properties (43) and (44).

8. Conclusions

In this paper, we considered a nonlocal boundary value problem T_ω for a weak nonlinear partial differential equation of mixed type with fractional Hilfer operator $D^{\alpha,\gamma}$ in a positive rectangular domain $\Omega_1 = \{0 < t < b, 0 < x, y < l\}$ and with spectral parameter ω in a negative rectangular domain $\Omega_2 = \{-a < t < 0, 0 < x, y < l\}$.

The set of positive solutions $\Im$ of trigonometric Equation (33) with respect to spectral parameter ω was called a set of irregular values of the spectral parameter ω. The set of the remaining values of the spectral parameter $\aleph = (0; \infty) \setminus \Im$ was called a set of regular values of the spectral parameter ω.

For all regular values of the spectral parameter ω the quantity $\Delta_{n,m}(\omega)$ was nonzero. So, for large n, m the values of $\Delta_{n,m}(\omega)$ could not become quite small and there the problem of "small denominators" did not arise. Therefore, for regular values of the spectral parameter ω the quantity $\Delta_{n,m}(\omega)$ was separated from zero and we considered the questions of one value solvability of the considering boundary value problems (1)–(5).

We studied the boundary value problem T_ω with following assumptions:

$$\varphi(x,y) \in C^3[0;l]^2, \varphi_{xxxx}(x,y) \in L_2[0;l]^2, \varphi_{yyyy}(x,y) \in L_2[0;l]^2;$$

$$f_i(x,y,u) \in C^{3,3,0}_{x,y,u}\left([0;l]^2 \times \mathbb{R}\right), f_{ixxxx}(x,y,u) \in L_2\left([0;l]^2 \times \mathbb{R}\right),$$

$$f_{iyyyy}(x,y,u) \in L_2\left([0;l]^2 \times \mathbb{R}\right);$$

$$\chi_{11} = \max_{i=\overline{1,3}} \max_{n,m\in\mathbb{N}} \max_{t\in[0;b]} \left| t^{1-\gamma} \eta_{inm}(t,\omega) \right| < \infty; \chi_{21} = \max_{i=\overline{1,3}} \max_{n,m\in\mathbb{N}} \max_{t\in[-a;0]} |\xi_{inm}(t,\omega)| < \infty;$$

$$\chi_{30} = \|\varphi_{xxxxyyyy}(x,y)\|_{L_2[0;l]^2} < \infty; \chi_{3i} = \|f_{ixxxxyyyy}(x,y,\gamma)\|_{L_2[0;l]^2} < \infty;$$

$$|f_{ixxxxyyyy}(x,y,\gamma_1) - f_{ixxxxyyyy}(x,y,\gamma_2)| \le K_i(x,y)|\gamma_1 - \gamma_2|,$$

$$K_{0i} = \|K_i(x,y)\|_{L_2[0;l]^2} < \infty;$$

$$|\Theta_i(\xi,x,y,u_1) - \Theta_i(\xi,x,y,u_2)| \le \Theta_{1i}(x,y)|u_1 - u_2|,$$

$$\Theta_{2i} = \|\Theta_{1i}(x,y)\|_{L_2[0;l]^2} < \infty, i=1,2;$$

$$\rho = \gamma_2(\gamma_1 + \gamma_3)\gamma_4 < 1, \gamma_4 = \max\{bK_{01}\Theta_{21}; aK_{02}\Theta_{22}\}.$$

If these conditions were fulfilled, then the boundary value problem T_ω was uniquely solvable for regular values of the spectral parameter $\omega \in \aleph$ with these solutions represented in the form of the Fourier series (53) and (54) in the domains Ω_1 and Ω_2, respectively. There the series (53), (54) and (57)–(62) were convergent absolutely and uniformly in the corresponding domains Ω_1 or Ω_2.

For irregular values of the spectral parameter $\omega \in \Im$ and for some k, $s = k_1, ..., k_s$ the problem T_ω had an infinite number of solutions in the form of series (64) and (65), if there the condition (63) was fulfilled.

Author Contributions: Conceptualization, T.K.Y. and B.J.K. All authors have read and agreed to the published version of the manuscript.

Funding: This research received no external funding.

Conflicts of Interest: The author declares no conflicts of interest.

References

1. *Handbook of Fractional Calculus with Applications*; Tenreiro Machado, J.A., Ed.; Walter de Gruyter GmbH: Berlin, Germany, 2019; Volumes 8, pp. 47–85.
2. Sun, H.; Chang, A.; Zhang, Y.; Chen, W. A review on variable-order fractional differential equations: Mathematical foundations, physical models, numerical methods and applications. *Fract. Calc. Appl. Anal.* **2019**, *22*, 27–59. [CrossRef]
3. Kumar, D.; Baleanu, D. Fractional Calculus and Its Applications in Physics. *Front. Phys.* **2019**, *7*. [CrossRef]
4. Saxena Ram, K.; Garra, R.; Orsingher, E. Analytical solution of space-time fractional telegraph-type equations involving Hilfer and Hadamard derivatives. *Integral Transform. Spec. Funct.* **2016**, *27*, 30–42. [CrossRef]
5. Patnaik, S.; Hollkamp, J.P.; Semperlotti, F. Applications of variable-order fractional operators: A review. *Proc. R. Soc.* **2020**. [CrossRef] [PubMed]
6. Klafter, J.; Lim, S.C.; Metzler, R. *Fractional Dynamics, Recent Advances*; World Scientific: Singapore, 2011; Chapter 9.
7. Xu, C.; Yu, Y.; Chen, Y.Q.; Lu, Z. Forecast analysis of the epidemic trend of COVID-19 in the United States by a generalized fractional-order SEIR model. *arXiV* **2020**. . [CrossRef]
8. Hilfer, R. *Application of Fractional Calculus in Physics*; World Scientific Publishing Company: Singapore, 2000.
9. Hilfer, R. Experimental evidence for fractional time evolution in glass forming materials. *Chem. Phys.* **2002**, *284*, 399–408. [CrossRef]
10. Hilfer, R. On fractional relaxation. *Fractals* **2003**, *11*, 251–257. [CrossRef]
11. Sandev, T.; Metzler, R.; Tomovski, Ž. Fractional diffusion equation with a generalized Riemann–Liouville time fractional derivative. *J. Phys. A Math. Theor.* **2011**, *44*, 255203. [CrossRef]
12. Tomovski, Ž.; Sandev, T.; Metzler, R.; Dubbeldam, J. Generalized space-time fractional diffusion equation with composite fractional time derivative. *Phys. A* **2012**, *391*, 2527–2542. [CrossRef]
13. Garra, R.; Gorenflo, R.; Polito, F.; Tomovski, Ž. Hilfer-Prabhakar derivatives and some applications. *Appl. Math. Comput.* **2014**, *242*, 576–589. [CrossRef]
14. Hilfer, R.; Luchko, Y.; Tomovski, Ž. Operational method for the solution of fractional differential equations with generalized Riemann–Liouville fractional derivatives. *Fract. Calc. Appl. Anal.* **2009**, *12*, 299–318.
15. Myong-Ha, K.; Guk-Chol, R.; Hyong-Chol, O. Operational method for solving multi-term fractional differential equations with the generalized fractional derivatives. *Fract. Calc. Appl. Anal.* **2014**, *17*, 79–95.
16. Sandev, T.; Tomovski, Ž. *Fractional Equations and Models: Theory and Applications*; Springer Nature Switzerland AG: Cham, Switzerland, 2019.
17. Al-Ghafri, K.S.; Rezazadeh, H. Solitons and other solutions of (3+1)-dimensional space-time fractional modified KdV-Zakharov-Kuznetsov equation. *Appl. Math. Nonlinear Sci.* **2019**, *4*, 289–304. [CrossRef]
18. Delbosco, D.; Rodino, L. Existence and uniqueness for a nonlinear fractional differential equation. *J. Math. Anal. Appl.* **1996**, *204*, 609–625. [CrossRef]
19. Ding, X.; Ahmad, B. Analytical solutions to fractional evolution equations with almost sectorial operators. *Adv. Differ. Equ.* **2016**, *2016*, 1–25. [CrossRef]
20. Furati, K.M.; Kassim, M.D.; Tatar, N.E. Existence and uniqueness for a problem involving Hilfer fractional derivative. *Comput. Math. Appl.* **2012**, *64*, 1616–1626. [CrossRef]
21. He, J.H. Some applications of nonlinear fractional differential equations and their approximations. *Bull. Sci. Technol.* **1999**, *15*, 86–90.
22. Jaiswal, A.; Bahuguna, D. Hilfer Fractional Differential Equations with Almost Sectorial Operators. *Differ Equ. Dynam. Syst.* **2020**, *13*, 18. [CrossRef]
23. Partohaghighi, M.; Inc, M.; Bayram, M.; Baleanu, D. On Numerical Solution Of The Time Fractional Advection-Diffusion Equation Involving Atangana-Baleanu-Caputo Derivative. *Open Phys.* **2019**, *17*, 816–822. [CrossRef]
24. Tripathi, B.; Sharma, B.; Sharma, M. Modeling and analysis of MHD two-phase blood flow through a stenosed artery having temperature-dependent viscosity. *Eur. Phys. J. Plus.* **2019** *134*, 1–17. [CrossRef]
25. Zhou, Y. *Basic Theory of Fractional Differential Equations*; World Scientific: Singapore, 2014.
26. Arqub, O.A.; Al-Smadi, M. Atangana–Baleanu fractional approach to the solutions of Bagley–Torvik and Painlevé equations in Hilbert space. *Chaos Solitons Fractals* **2018**, *117*, 161–167. [CrossRef]

27. Malik, S.A.; Aziz, S. An inverse source problem for a two parameter anomalous diffusion equation with nonlocal boundary conditions. *Comput. Math. Appl.* **2017**, *73*, 12. [CrossRef]
28. Aziz, S.; Malik, S.A. Identification of an unknown source term for a time fractional fourth-order parabolic equation. *Electron. J. Differ. Equ.* **2016**, *2016*, 1–20.
29. Sabitov, K.B.; Safin, E.M. The inverse problem for a mixed-type parabolic-hyperbolic equation in a rectangular domain. *Russ. Math.* **2010**, *54*, 48–54. [CrossRef]
30. Sabitov, K.B.; Martem'yanova, N.V. A nonlocal inverse problem for a mixed-type equation. *Russ. Math.* **2011**, *55*, 61–74. [CrossRef]
31. Sabitov, K.B.; Sidorov, S.N. On a nonlocal problem for a degenerating parabolic-hyperbolic equation. *Differ. Equ.* **2014**, *50*, 352–361. [CrossRef]
32. Sabitov, K.B. *On the Theory of Mixed Type Equations*; Fizmatlit Publ. House: Moscow, Russia, 2014; 301p. (In Russian)
33. Urinov, A.K.; Nishonova, S.T. A problem with integral conditions for an elliptic-parabolic equation. *Math. Notes* **2017**, *102*, 68–80. [CrossRef]
34. Yuldashev, T.K. Solvability of a boundary value problem for a differential equation of the Boussinesq type. *Differ. Equ.* **2018**, *54*, 1384–1393. [CrossRef]
35. Yuldashev, T.K. Nonlocal inverse problem for a pseudohyperbolic-pseudoelliptic type integro-differential equations. *Axioms* **2020**, *9*, 45. [CrossRef]
36. Zikirov, O.S. A non-local boundary value problem for third-order linear partial differential equation of composite type. *Math. Model. Anal.* **2009**, *14*, 407–421. [CrossRef]
37. Yuldashev, T.K. Mixed value problem for a nonlinear differential equation of fourth order with small parameter on the parabolic operator. *Comput. Math. Math. Phys.* **2011**, *51*, 1596–1604. [CrossRef]
38. Yuldashev, T.K. Mixed value problem for nonlinear integro-differential equation with parabolic operator of higher power. *Comput. Math. Math. Phys.* **2012**, *52*, 105–116. [CrossRef]
39. Yuldashev, T.K. Inverse problem for a nonlinear Benney–Luke type integro-differential equations with degenerate kernel. *Russ. Math.* **2016**, *60*, 53–60. [CrossRef]
40. Yuldashev, T.K. Nonlocal mixed-value problem for a Boussinesq-type integrodifferential equation with degenerate kernel. *Ukr. Math. J.* **2016**, *68*, 1278–1296. [CrossRef]
41. Yuldashev, T.K. Mixed problem for pseudoparabolic integrodifferential equation with degenerate kernel. *Differ. Equ.* **2017**, *53*, 99–108. [CrossRef]
42. Yuldashev, T.K. On a boundary-value problem for Boussinesq type nonlinear integro-differential equation with reflecting argument. *Lobachevskii J. Math.* **2020**, *41*, 111–123. [CrossRef]
43. Berdyshev, A.S.; Kadirkulov, B.J. On a Nonlocal Problem for a Fourth-Order Parabolic Equation with the Fractional Dzhrbashyan-Nersesyan Operator. *Differ. Equ.* **2016**, *52*, 122–127. [CrossRef]
44. Berdyshev, A.S.; Cabada, A.; Kadirkulov, B.J. The Samarskii-Ionkin type problem for fourth order parabolic equation with fractional differential operator. *Comput. Math. Appl.* **2011**, *62*, 3884–3893. [CrossRef]
45. Kerbal, S.; Kadirkulov, B.J.; Kirane, M. Direct and inverse problems for a Samarskii-Ionkin type problem for a two dimensional fractional parabolic equation. *Progr. Fract. Differ. Appl.* **2018**, *4*, 1–14. [CrossRef]

© 2020 by the authors. Licensee MDPI, Basel, Switzerland. This article is an open access article distributed under the terms and conditions of the Creative Commons Attribution (CC BY) license (http://creativecommons.org/licenses/by/4.0/).

Article

A Rosenzweig–MacArthur Model with Continuous Threshold Harvesting in Predator Involving Fractional Derivatives with Power Law and Mittag–Leffler Kernel

Hasan S. Panigoro [1,2], Agus Suryanto [1,*], Wuryansari Muharini Kusumawinahyu [1] and Isnani Darti [1]

[1] Department of Mathematics, Faculty of Mathematics and Natural Sciences, University of Brawijaya, Malang 65145, Indonesia; hspanigoro@ung.ac.id (H.S.P.); wmuharini@ub.ac.id (W.M.K.); isnanidarti@ub.ac.id (I.D.)

[2] Department of Mathematics, Faculty of Mathematics and Natural Sciences, State University of Gorontalo, Bone Bolango 96119, Indonesia

* Correspondence: suryanto@ub.ac.id

Received: 17 September 2020; Accepted: 19 October 2020; Published: 22 October 2020

Abstract: The harvesting management is developed to protect the biological resources from over-exploitation such as harvesting and trapping. In this article, we consider a predator–prey interaction that follows the fractional-order Rosenzweig–MacArthur model where the predator is harvested obeying a threshold harvesting policy (THP). The THP is applied to maintain the existence of the population in the prey–predator mechanism. We first consider the Rosenzweig–MacArthur model using the Caputo fractional-order derivative (that is, the operator with the power-law kernel) and perform some dynamical analysis such as the existence and uniqueness, non-negativity, boundedness, local stability, global stability, and the existence of Hopf bifurcation. We then reconsider the same model involving the Atangana–Baleanu fractional derivative with the Mittag–Leffler kernel in the Caputo sense (ABC). The existence and uniqueness of the solution of the model with ABC operator are established. We also explore the dynamics of the model with both fractional derivative operators numerically and confirm the theoretical findings. In particular, it is shown that models with both Caputo operator and ABC operator undergo a Hopf bifurcation that can be controlled by the conversion rate of consumed prey into the predator birth rate or by the order of fractional derivative. However, the bifurcation point of the model with the Caputo operator is different from that of the model with the ABC operator.

Keywords: Rosenzweig–MacArthur model; fractional derivatives; threshold harvesting

1. Introduction

More than 50 years after the model has been proposed, the Rosenzweig–MacArthur predator–prey model [1] has been consistently developed by many scholars to approach the real world phenomena with more realistic mathematical models. The commonsensical modified Rosenzweig–MacArthur models are accomplishable by considering the biological perspectives of ecosystem conditions, for instance the stage structure [2,3], the refuge effect [4–8], the fear effect [9], the Allee effect [10,11], the intraspecific competition [12,13], the cannibalism [14], the infectious disease [15–17], and so forth.

On the other hand, the modeling also contemplates the optimal management of bioeconomic resources as in fishery and pest management. Some researchers put an intervention into the predator–prey model such as the harvesting to one or more population [8,18–22]. To protect the population from over-exploitation during the harvesting, some management techniques have been established. One of the famous technique is a continuous threshold harvesting policy (THP) (see [23–29]). THP is regulated as follows: when the population density above the threshold level, harvesting is permitted; when the population density drops below the threshold level, harvesting is prohibited.

In 2013, Lv et al. [26] proposed the following Rosenzweig–MacArthur model with THP in predator

$$\begin{aligned}\frac{dx}{dt} &= rx\left(1-\frac{x}{K}\right) - \frac{mxy}{a+x}, \\ \frac{dy}{dt} &= \frac{nxy}{a+x} - dy - H(y),\end{aligned} \quad (1)$$

where

$$H(y) = \begin{cases} 0 & , \text{ if } y < T, \\ \frac{h(y-T)}{c+(y-T)} & , \text{ if } y \geq T. \end{cases}$$

We describe the biological interpretation of variables and parameters of model (1) in Table 1. Model (1) represents an interaction between two populations with a prey–predator relationship, where THP is only applied for the predator to preserving its populations. Some appealing examples of the ecological model (1) are given by the interaction between *Sycanus* sp. and *Setothosea asigna* and between *Rhinocoris* sp. and *Spodoptera litura*. Shepard [30] reported that *Sycanus* sp. and *Rhinocoris* sp. are the natural predators of the pests such as *Setothosea asigna* and *Spodoptera litura* which exist in agricultural lands and plantations. The worrying problem is: How if the density of insects such as *Sycanus* sp. and *Rhinocoris* sp. uncontrolled? One solution is applying THP as in model (1).

Table 1. Description of variables and parameters of the model (1).

Variables and Parameters	Description
$x(t)$	The density of prey
$y(t)$	The density of predator
r	The intrinsic growth rate of prey
K	The environmental carrying capacity of prey
m	The maximum uptake rate for prey
n	The conversion rate of consumed prey into predator birth
a	The environment protection for prey
d	The natural death rate of predator
h	The harvesting rate
c	The half saturation constant for harvesting
T	The threshold level of harvesting

Lv et al. [26] successfully explored the dynamics of the model (1) including the local stability and the existence of various phenomena (saddle-node, Hopf, cusp, and Bogdanov–Taken bifurcations). Despite their success works, the model with the first-order derivative is limited to its capability of involving all previous conditions to the growth rates of both predator and prey. The growth rates of both populations in the model (1) depend only on the current state. Biologically, the growth rates must be dependent on all of the previous conditions which are known as the memory effects. To account for such memory effects, some researchers proposed to apply the fractional-order derivative instead of the first-order derivative when expressing the growth rate of the population. The fractional-order models are naturally related to systems with memory which exists in most biological models [7,31]. The fractional-order models are also well-liked due to their capability in providing an exact description of different nonlinear phenomena [32]. In recent years, the development of fractional-order models grows rapidly and becomes popular in studying the dynamical behavior of predator–prey interaction [17,33–38]. It has been shown that the order of the fractional derivate significantly affects the dynamical behavior of the models, which is in contrast to the first-order derivative models that depend only on the values of parameters.

In this paper, we modify the model of Lv et al. [26] by applying the fractional-order derivative to the left-hand sides of the first-order differential Equation (1). We use two types of fractional-order derivatives, namely the Caputo operator (that is, the operator with the power law kernel) [39] and Atangana–Baleanu operator [40]. The basic difference between these two operators lies on their kernel. Atangana–Baleanu operator has a non-singular and non-local (that is, Mittag–Leffler) kernel while the Caputo operator does not [41,42]. From the biological meanings, a model with Atangana–Baleanu operator gives better results in applying memory effects [43–45]. Nevertheless, the Caputo operator has more complex analytical tools in investigating the dynamics of the model such as the local stability [46–50], the global stability [51,52], bifurcation theory [52–54], and so on. By considering the deficiencies and advantages, the model with Caputo and Atangana–Baleanu operator are employed in our work. Based on our literature review, the dynamics of the model (1) with Caputo and Atangana–Baleanu operator have never been studied. For this reason, we are interested in investigating the dynamical behavior of model (1) using both Caputo and Atangana–Baleanu fractional-order operators.

If the first-order derivatives $\frac{d}{dt}$ at the left hand sides of model (1) are replaced by the fractional-order derivatives $\mathcal{D}_t^\alpha$, then we obtain

$$\mathcal{D}_t^\alpha x = \hat{r} x \left(1 - \frac{x}{K}\right) - \frac{\hat{m} x y}{a + x},$$
$$\mathcal{D}_t^\alpha y = \frac{\hat{n} x y}{a + x} - \hat{d} y - H(y). \tag{2}$$

Note that the left hand sides of model (2) have the dimension of $(\text{time})^{-\alpha}$, while the parameters at the right hand sides of model (2) such as $\hat{r}, \hat{m}, \hat{n}, \hat{d}$, and $\hat{h}$ have the dimension of $(\text{time})^{-1}$; this shows the inconsistency of physical dimensions in the model (2) (see [55,56]). To overcome such inconsistency, we rescale the parameters in the model (2) to get the following model

$$\mathcal{D}_t^\alpha x = \hat{r}^\alpha x \left(1 - \frac{x}{K}\right) - \frac{\hat{m}^\alpha x y}{a + x},$$
$$\mathcal{D}_t^\alpha y = \frac{\hat{n}^\alpha x y}{a + x} - \hat{d}^\alpha y - H(y), \tag{3}$$

where

$$H(y) = \begin{cases} 0, & \text{if } y < T, \\ \frac{\hat{h}^\alpha (y - T)}{c + (y - T)}, & \text{if } y \geq T. \end{cases}$$

By applying new parameters $r = \hat{r}^\alpha, m = \hat{m}^\alpha, n = \hat{n}^\alpha, d = \hat{d}^\alpha$, and $h = \hat{h}^\alpha$, we obtain

$$\mathcal{D}_t^\alpha x = r x \left(1 - \frac{x}{K}\right) - \frac{m x y}{a + x},$$
$$\mathcal{D}_t^\alpha y = \frac{x y}{a + x} - d y - H(y), \tag{4}$$

where

$$H(y) = \begin{cases} 0, & \text{if } y < T, \\ \frac{h(y - T)}{c + (y - T)}, & \text{if } y \geq T. \end{cases}$$

This paper is organized as follows. In Section 2, we investigate the dynamics of model (4) with the Caputo operator. We identify the existence, uniqueness, non-negativity, and boundedness of solutions. Furthermore, we explore the dynamics of the model by examining the existence of the equilibrium points, their local and global stability, and the existence of Hopf bifurcation. In Section 3, the existence and uniqueness of solutions of the model with Atangana–Baleanu operator are verified. In Section 4, we explore the dynamics of the model using both operators numerically. We demonstrate numerically the stability of the equilibrium point, and the occurrence of forward and Hopf bifurcations. We end our works with a brief conclusion in Section 5.

2. The Caputo Fractional-Order Rosenzweig–MacArthur Model with THP in Predator

2.1. Model Formulation

The operator of Caputo fractional-order derivative is defined as follows

Definition 1. *[48] Let $\alpha \in (0,1]$, $f \in C^n([0,+\infty), \mathbb{R})$, and $\Gamma(\cdot)$ is the Gamma function. The Caputo fractional derivative of order-α is defined by*

$$^C\mathcal{D}_t^\alpha f(t) = \frac{1}{\Gamma(1-\alpha)} \int_0^t (t-s)^{-\alpha} f'(s) ds, t \geq 0. \tag{5}$$

The kernel of Caputo operator is known as the power law kernel. By applying Definition 1 to model (4), we get the Caputo fractional order Rosenzweig–MacArthur model with THP in predator

$$\begin{aligned} ^C\mathcal{D}_t^\alpha x &= rx\left(1 - \frac{x}{K}\right) - \frac{mxy}{a+x} \equiv F_1, \\ ^C\mathcal{D}_t^\alpha y &= \frac{nxy}{a+x} - dy - H(y) \equiv F_2. \end{aligned} \tag{6}$$

2.2. Existence and Uniqueness

In this part, we study the existence and uniqueness of model (6).

Lemma 1. *[57] Consider a Caputo fractional-order system*

$$^C\mathcal{D}_t^\alpha x(t) = f(t, x(t)), t > 0, x(0) \geq 0, \alpha \in (0,1], \tag{7}$$

where $f : (0, \infty) \times \Theta \to \mathbb{R}^n$, $\Theta \subseteq \mathbb{R}^n$. A unique solution of Equation (7) on $(0, \infty) \times \Theta$ exists if $f(t, x(t))$ satisfies the locally Lipschitz condition with respect to x.

Since the right hand-side of model (6) is a piecewise function which is switched when the number of predators passes through the threshold level, we divide the existence and uniqueness of the solution into two cases, namely $y \geq T$ and $y < T$. We start from $y \geq T$. Consider the region $\Theta \times [0, T_+]$ where $\Theta := \{(x,y) \in \mathbb{R}^2 : \max(|x|, |y|) \leq \gamma, y \geq T\}$, $T_+ < +\infty$, and a mapping $F(\Lambda) = (F_1(\Lambda), F_2(\Lambda))$. For any $\Lambda = (x, y) \in \Theta$ and $\bar{\Lambda} = (\bar{x}, \bar{y}) \in \Theta$, we obtain

$$\|F(\Lambda) - F(\bar{\Lambda})\| = |F_1(\Lambda) - F_1(\bar{\Lambda})| + |F_2(\Lambda) - F_2(\bar{\Lambda})|$$

$$= \left| rx\left(1 - \frac{x}{K}\right) - \frac{mxy}{a+x} - \left(r\bar{x}\left(1 - \frac{\bar{x}}{K}\right) - \frac{m\bar{x}\bar{y}}{a+\bar{x}}\right) \right| +$$

$$\left| \frac{nxy}{a+x} - dy - H(y) - \left(\frac{n\bar{x}\bar{y}}{a+\bar{x}} - d\bar{y} - H(\bar{y})\right) \right|$$

$$= \left| r(x-\bar{x}) - \frac{r}{K}\left(x^2 - \bar{x}^2\right) - m\left(\frac{xy}{a+x} - \frac{\bar{x}\bar{y}}{a+\bar{x}}\right) \right| +$$

$$\left| n\left(\frac{xy}{a+x} - \frac{\bar{x}\bar{y}}{a+\bar{x}}\right) - d(y-\bar{y}) - (H(y) - H(\bar{y})) \right|$$

$$= r|x-\bar{x}| + \frac{r}{K}|x+\bar{x}||x-\bar{x}| + (m+n)\left|\frac{xy}{a+x} - \frac{\bar{x}\bar{y}}{a+\bar{x}}\right| +$$

$$d|y-\bar{y}| + \left|\frac{h(y-T)}{c+(y-T)} - \frac{h(\bar{y}-T)}{c+(\bar{y}-T)}\right|$$

$$= r|x-\bar{x}| + \frac{r}{K}|x+\bar{x}||x-\bar{x}| + (m+n)\left|\frac{ay(x-\bar{x}) + (a\bar{x}+\bar{x}x)(y-\bar{y})}{(a+x)(a+\bar{x})}\right| +$$

$$d|y-\bar{y}| + \left|\frac{ch(y-\bar{y})}{(c+y-T)(c+\bar{y}-T)}\right|$$

$$\leq r|x-\bar{x}| + \frac{2\gamma r}{K}|x-\bar{x}| + \frac{m+n}{a^2}|ay(x-\bar{x}) + (a\bar{x}+\bar{x}x)(y-\bar{y})| +$$

$$d|y-\bar{y}| + \frac{h}{c}|y-\bar{y}|$$

$$\leq r|x-\bar{x}| + \frac{2\gamma r}{K}|x-\bar{x}| + \frac{(m+n)\gamma}{a}|x-\bar{x}| +$$

$$\frac{(m+n)(a\gamma+\gamma^2)}{a^2}|y-\bar{y}| + d|y-\bar{y}| + \frac{h}{c}|y-\bar{y}|$$

$$= \left(r + \frac{2\gamma r}{K} + \frac{(m+n)\gamma}{a}\right)|x-\bar{x}| + \left(\frac{(m+n)(a\gamma+\gamma^2)}{a^2} + d + \frac{h}{c}\right)|y-\bar{y}|$$

$$\leq M_1 \|\Lambda - \bar{\Lambda}\|,$$

where $M_1 = \max\left\{r + \frac{2\gamma r}{K} + \frac{(m+n)\gamma}{a}, \frac{(m+n)(a\gamma+\gamma^2)}{a^2} + d + \frac{h}{c}\right\}$. Hence, $F(\Lambda)$ satisfies the Lipschitz condition for $y \geq T$. By using similar approaches, when $y < T$, it is easy to check that $\|F(\Lambda) - F(\bar{\Lambda})\| \leq M_2 \|\Lambda - \bar{\Lambda}\|$, where $M_2 = \max\left\{r + \frac{2\gamma r}{K} + \frac{(m+n)\gamma}{a}, \frac{(m+n)(a\gamma+\gamma^2)}{a^2} + d\right\}$ and hence the Lipschitz condition is also satisfied. According to Lemma 1, model (6) with non-negative initial condition has a unique solution $\Lambda(t) = (x(t), y(t)) \in \Theta$. Thus, we establish the following theorem.

Theorem 1. *For each non-negative initial condition* $(x_0, y_0) \in \Theta$, *there exists a unique solution* $(x(t), y(t)) \in \Theta$ *of model* (6), *which is defined for all* $t \geq 0$.

2.3. Non-Negativity and Boundedness

The solution of model (6) is required to be nonnegative and bounded to establish a biologically well-behaved model. To determine the non-negativity and boundedness of the solution of model (6), the following lemmas are needed.

Lemma 2. *[58] Let* $0 < \alpha \leq 1$. *Suppose that* $f(t) \in C[a,b]$ *and* $^C D_t^\alpha f(t) \in C[a,b]$. *If* $^C D_t^\alpha f(t) \geq 0, \forall t \in (a,b)$, *then* $f(t)$ *is a non-decreasing function for each* $t \in [a,b]$. *If* $^C D_t^\alpha f(t) \leq 0, \forall t \in (a,b)$, *then* $f(t)$ *is a non-increasing function for each* $t \in [a,b]$.

Lemma 3. *(Standard comparison theorem for Caputo fractional-order derivative [31]). Let $x(t) \in C([0, +\infty))$. If $x(t)$ satisfies ${}^C\mathcal{D}_t^\alpha x(t) + \lambda x(t) \leq \mu, x(0) = x_0$, where $\alpha \in (0, 1], (\lambda, \mu) \in \mathbb{R}^2$ and $\lambda \neq 0$, then $x(t) \leq \left(x_0 - \frac{\mu}{\lambda}\right) E_\alpha[-\lambda t^\alpha] + \frac{\mu}{\lambda}$.*

In the following theorem, we prove the non-negativity and boundedness of solutions using the above lemmas.

Theorem 2. *All solutions of model (6) with non-negative initial conditions are non-negative and uniformly bounded.*

Proof. We start by proving that, if the initial condition is non-negative, then $x(t) \geq 0$ for all $t \to \infty$. Suppose that it is incorrect; then, we can find $t_1 > 0$ such that

$$\begin{cases} x(t) & > \quad 0, 0 \leq t < t_1, \\ x(t_1) & = \quad 0, \\ x(t_1^+) & < \quad 0 \end{cases} \tag{8}$$

By employing (8) and the first equation of model (6), we obtain

$${}^C\mathcal{D}_t^\alpha x(t_1)\Big|_{x(t_1)=0} = 0. \tag{9}$$

Based on Lemma 2, we have $x(t_1^+) = 0$, which contradicts the fact that $x(t_1^+) < 0$. Thus, $x(t) \geq 0$ for all $t \geq 0$. Using a similar procedure, we conclude $y(t) \geq 0$ for all $t > 0$.

Now, we adopt the similar manner as in [34] to show the boundedness of solutions. By setting up a function $\mathcal{V}(t) = x + \frac{my}{n}$, we get

$$\begin{aligned}
{}^C\mathcal{D}_t^\alpha \mathcal{V}(t) + d\mathcal{V}(t) &= {}^C\mathcal{D}_t^\alpha x + \frac{m}{n}{}^C\mathcal{D}_t^\alpha y + dx + \frac{dmy}{n} \\
&= rx\left(1 - \frac{x}{K}\right) - \frac{mxy}{a+x} + \frac{m}{n}\left(\frac{nxy}{a+x} - dy - H(y)\right) + dx + \frac{dmy}{n} \\
&= (d+r)x - \frac{rx^2}{K} - \frac{mH(y)}{n} \\
&= -\frac{r}{K}\left(x - \frac{(d+r)K}{2r}\right)^2 + \frac{(d+r)^2 K}{4r} - \frac{mH(y)}{n} \\
&\leq \frac{(d+r)^2 K}{4r}.
\end{aligned}$$

According to the standard comparison theorem for Caputo fractional-order derivative in Lemma 3, we achieve the following inequality

$$\mathcal{V}(t) \leq \left(\mathcal{V}(0) - \frac{(d+r)^2 K}{4r}\right) E_\alpha[-d(t)^\alpha] + \frac{(d+r)^2 K}{4r},$$

where E_α is the one-parameter Mittag–Leffler function. Since $E_\alpha[-d(t)^\alpha] \to 0$ as $t \to 0$, we acquire $\mathcal{V}(t) \to \frac{(d+r)^2 K}{4r}$ for $t \to \infty$. Hence, with non-negative initial condition, all solutions are restricted to the region Θ_M where

$$\Theta_M := \left\{(x,y) \in \mathbb{R}_+^2 : x + \frac{my}{n} \leq \frac{(d+r)^2 K}{4r} + \varepsilon, \varepsilon > 0\right\}.$$

Consequently, all solutions of model (6) are uniformly bounded. □

2.4. The Existence of Equilibrium Point

The first commonplace technique in studying the dynamical behavior of a fractional-order system is investigating the existence of the equilibrium point. Due to the biological nature, we give the following definition.

Definition 2. *Consider a Caputo fractional-order system*

$$^C D_t^\alpha \vec{x} = \vec{f}(\vec{x}); \vec{x}(0) = \vec{x}_0; \alpha \in (0,1]. \tag{10}$$

A point $\vec{x}^*$ is called an equilibrium point of Equation (10) if it satisfies $\vec{f}(\vec{x}^*) = 0$. Particularly, it is called the biological equilibrium point of Equation (10) if it satisfies $\vec{x}^* \geq 0$.

Based on Definition 2, the equilibrium point of model (6) is obtained by solving

$$\left[r - \frac{rx}{K} - \frac{my}{a+x} \right] x = 0,$$
$$\frac{nxy}{a+x} - dy - H(y) = 0. \tag{11}$$

Thus, we get four feasible biological equilibrium points as follows.

(i) The equilibrium points when $y < T$ are
  - (i.a) the origin point $E_0 = (0,0)$ which always exists;
  - (i.b) the predator extinction point $E_1 = (K,0)$ which always exists; and
  - (i.c) the first co-existence point $\hat{E} = \left(\hat{x}, \frac{(K-\hat{x})(a+\hat{x})r}{mK} \right)$, with $\hat{x} = \frac{ad}{n-d}$, which exists if $n > \frac{ad}{K} + d$ and $(K - \hat{x})(a + \hat{x}) < \frac{TmK}{r}$.

(ii) The equilibrium point when $y \geq T$ is the second co-existence point $E^* = (x^*, y^*)$ where $y^* = \frac{(K-x^*)(a+x^*)r}{mK}$ and x^* is the positive roots of polynomial $\beta_1 x^4 + \beta_2 x^3 + \beta_3 x^2 + \beta_4 x + \beta_5 = 0$ where

$$\beta_1 = (n-d)r^2,$$
$$\beta_2 = [(an + 2dK) - 2(ad + nK)]r^2,$$
$$\beta_3 = (nrK + 4adr + cdm + mnT + hm)rK$$
$$\quad - ((drK + 2anr + cmn + dmT)K + a^2 dr)r,$$
$$\beta_4 = ((anr + cmn + dmT)K + (2adr + hm + cdm)a)rK$$
$$\quad - ((2adr + hm + cdm + mnT)K + admT)rK,$$
$$\beta_5 = [(adr + hm)mT - (adr + hm + cdm)ar]K^2.$$

E^* exists if $0 < x^* < K$ and $(K - x^*)(a + x^*) \geq \frac{TmK}{r}$.

2.5. Local Asymptotic Stability

In this part, we discuss the local stability of E_0, E_1, $\hat{E}$, and E^*. For this aim, we need the following theorem.

Theorem 3. *(Matignon condition [48,59]) The equilibrium point $\vec{x}^*$ of system (10) is locally asymptotically stable if all eigenvalues λ_j of the Jacobian matrix $J = \partial \vec{f} / \partial \vec{x}$ evaluated at $\vec{x}^*$ satisfy $|\arg(\lambda_j)| > \alpha \pi / 2$. If there exists at least one eigenvalue satisfying $|\arg(\lambda_k)| > \alpha \pi / 2$ and $|\arg(\lambda_l)| < \alpha \pi / 2$, $k \neq l$, then $\vec{x}^*$ is a saddle-point.*

Now, we present Theorems 4–7 to show the local stability properties of E_0, E_1, $\hat{E}$, and E^*.

Theorem 4. *The origin point $E_0 = (0,0)$ is always a saddle point.*

Proof. When $E_0 = (0,0)$, the model (6) has Jacobian matrix $J(E_0) = \begin{bmatrix} r & 0 \\ 0 & -d \end{bmatrix}$, where its eigenvalues are $\lambda_1 = r > 0$ and $\lambda_2 = -d < 0$. It is clear that $|\arg(\lambda_1)| = 0 < \alpha\pi/2$ and $|\arg(\lambda_2)| = \pi > \alpha\pi/2$. Therefore, based on Theorem 3, E_0 is always a saddle point. □

Theorem 5. *The predator extinction point $E_1 = (K,0)$ is locally asymptotically stable if $n < \frac{ad}{K} + d$. Otherwise, if $n > \frac{ad}{K} + d$, then $E_1 = (K,0)$ is a saddle point.*

Proof. The Jacobian matrix of model (6) evaluated at E_1 is $J(E_1) = \begin{bmatrix} -r & -\frac{mK}{a+K} \\ 0 & \frac{nK}{a+K} - d \end{bmatrix}$. The eigenvalues of $J(E_1)$ are $\lambda_1 = -r < 0$ and $\lambda_2 = \frac{nK}{a+K} - d$. Clearly, $|\arg(\lambda_1)| = \pi > \alpha\pi/2$ and $|\arg(\lambda_2)| = \pi > \alpha\pi/2$ if $n < \frac{ad}{K} + d$ and $|\arg(\lambda_2)| = 0 < \alpha\pi/2$ if $n > \frac{ad}{K} + d$. Hence, we have the theorem. □

Remark 1. *It is noted that the existence condition for the first co-existence point $\hat{E}$ contradicts the stability condition of E_1. Consequently, if E_1 is locally asymptotically stable, then $\hat{E}$ does not exist. This condition also indicates the existence of forward bifurcation, which is confirmed numerically in the next section.*

Theorem 6. *Let $\Delta = \frac{4(K-\hat{x})anr\hat{x}}{(a+\hat{x})^2 K} - \left(\frac{K-\hat{x}}{a+\hat{x}} - 1\right)^2 \frac{r^2\hat{x}^2}{K^2}$ and $\hat{\alpha} = \frac{2}{\pi}\tan^{-1}\left(\frac{\sqrt{\Delta}(a+\hat{x})K}{(K-a-2\hat{x})r\hat{x}}\right)$. Suppose that one of the following statements is obeyed.*

(i) $\hat{x} > \frac{K-a}{2}$; or

(ii) $\hat{x} < \frac{K-a}{2}$, $\Delta > 0$, and $\alpha < \hat{\alpha}$.

Then, the first co-existence point $\hat{E} = \left(\hat{x}, \frac{(K-\hat{x})(a+\hat{x})r}{mK}\right)$ is locally asymptotically stable.

Proof. We first observe that the Jacobian matrix of model (6) evaluated at $\hat{E}$ is

$$J(\hat{E}) = \begin{bmatrix} \left(\frac{K-\hat{x}}{a+\hat{x}} - 1\right)\frac{r\hat{x}}{K} & -\frac{m\hat{x}}{a+\hat{x}} \\ \frac{(K-\hat{x})anr}{(a+\hat{x})mK} & 0 \end{bmatrix}. \qquad (12)$$

The eigenvalues of the Jacobian matrix (12) are the solutions of the characteristic equation

$$\lambda^2 - \left(\frac{K-\hat{x}}{a+\hat{x}} - 1\right)\frac{r\hat{x}}{K}\lambda + \frac{(K-\hat{x})anr\hat{x}}{(a+\hat{x})^2 K} = 0,$$

namely

$$\lambda_1 = \left(\frac{K-\hat{x}}{a+\hat{x}} - 1\right)\frac{r\hat{x}}{2K} + \frac{i\sqrt{\Delta}}{2},$$
$$\lambda_2 = \left(\frac{K-\hat{x}}{a+\hat{x}} - 1\right)\frac{r\hat{x}}{2K} - \frac{i\sqrt{\Delta}}{2}. \qquad (13)$$

When $\hat{x} > \dfrac{K-a}{2}$, the real parts of $\lambda_{1,2}$ are negative. Thus, the eigenvalues (13) always satisfy $|\arg(\lambda_{1,2})| > \alpha\pi/2$ for any Δ. If $\hat{x} < \dfrac{K-a}{2}$ and $\Delta \leq 0$, then $\lambda_1\lambda_2 = \dfrac{(K-\hat{x})anr\hat{x}}{(a+\hat{x})^2 K} > 0$ and $\lambda_1 + \lambda_2 = \left(\dfrac{K-\hat{x}}{a+\hat{x}} - 1\right)\dfrac{r\hat{x}}{K} > 0$, meaning that $|\arg(\lambda_{1,2})| = 0 < \alpha\pi/2$. If $\hat{x} < \dfrac{K-a}{2}$ and $\Delta > 0$, then $|\arg(\lambda_{1,2})| > \alpha\pi/2$ for $\alpha < \hat{\alpha}$. Hence, we prove the theorem. □

Theorem 7. *Let*

$$\xi = \dfrac{((y^*-T)^2 - cT)h}{y^*(c+y^*-T)^2} + \left(\dfrac{my^*}{(a+x^*)^2} - \dfrac{r}{K}\right)x^*,$$

$$\theta = \left(\dfrac{my^*}{(a+x^*)^2} - \dfrac{r}{K}\right)\dfrac{((y^*-T)^2 - cT)h}{y^*(c+y^*-T)^2}x^* + \left(1 - \dfrac{x^*}{a+x}\right)\dfrac{mnx^*y^*}{(a+x^*)^2}.$$

If one of the following statements is satisfied, then the second co-existence point $E^* = (x^*, y^*)$ is locally asymptotically stable:

(i) $\theta > 0$ and $\xi < 0$; or
(ii) $\xi^2 < 4\theta$, $\xi > 0$, and $\alpha < \alpha^*$.

Proof. The Jacobian matrix of model (6) evaluated at E^* is given by

$$J(E^*) = \begin{bmatrix} \left(\dfrac{my^*}{(a+x^*)^2} - \dfrac{r}{K}\right)x^* & -\dfrac{mx^*}{a+x^*} \\ \left(1 - \dfrac{x^*}{a+x^*}\right)\dfrac{ny^*}{a+x^*} & \dfrac{((y^*-T)^2 - cT)h}{y^*(c+y^*-T)^2} \end{bmatrix}. \quad (14)$$

The eigenvalues of (14) are obtained by solving the characteristic equation $\lambda^2 - \xi\lambda + \theta = 0$. Thus, we have $\lambda_{1,2} = \dfrac{\xi}{2} \pm \dfrac{\sqrt{\xi^2 - 4\theta}}{2}$. If $\theta > 0$ and $\xi < 0$, then $|\arg(\lambda_{1,2})| > \alpha\pi/2$. If $\xi^2 < 4\theta$ and $\xi > 0$, then $|\arg(\lambda_{1,2})| > \alpha\pi/2$ for $\alpha < \alpha^*$. Using Theorem 3, the local stability of E^* is completely proven. □

2.6. Global Asymptotic stability

To study the global stability of equilibrium points, we need the following lemmas.

Lemma 4. *[51] Let $x(t) \in C(\mathbb{R}_+)$, $x^* \in \mathbb{R}_+$, and its Caputo fractional derivative of order-α exists for any $\alpha \in (0,1]$. Then, for any $t > 0$, we have* ${}^C\mathcal{D}_t^\alpha\left[x(t) - x^* - x^*\ln\dfrac{x(t)}{x^*}\right] \leq \left(1 - \dfrac{x^*}{x(t)}\right){}^C\mathcal{D}_t^\alpha x(t)$.

Lemma 5. *(Generalized Lasalle Invariance Principle [52]). Suppose Ω is a bounded closed set and every solution of system*

$${}^C\mathcal{D}_t^\alpha x(t) = f(x(t)), \quad (15)$$

which starts from a point in Ω remains in Ω for all time. Let $V(x) : \Omega \to \mathbb{R}$ be a continuously differentiable function such that

$${}^C\mathcal{D}_t^\alpha V|_{(15)} \leq 0.$$

Let $E := \{x|{}^C\mathcal{D}_t^\alpha V|_{(15)} = 0\}$ and M be the largest invariant set of E. Then, every solution $x(t)$ originating in Ω tends to M as $t \to \infty$.

Since model (6) is basically a piecewise fractional-order differential equations that depends on T, the analysis of the global stability is split into two regions defined by $\Omega_1 := \{(x,y) : x \geq 0, y < T\}$ and $\Omega_2 := \{(x,y) : x \geq 0, y \geq T\}$. Therefore, the global stabilities of E_1, $\hat{E}$, and E^* are investigated as follows.

Theorem 8. If $n < \dfrac{ad}{K}$, then the predator extinction point $E_1 = (K, 0)$ is globally asymptotically stable in the region Ω_1.

Proof. By specifying a positive Lyapunov function $\mathcal{L}_1(x,y) = \left(x - K - K \ln \dfrac{x}{K}\right) + \dfrac{my}{n}$, and conforming to Lemma 4, we obtain

$$^C D_t^\alpha \mathcal{L}_1(x,y) \leq \left(\dfrac{x-K}{x}\right) {}^C D_t^\alpha x + \dfrac{m}{n} {}^C D_t^\alpha y$$

$$= (x - K)\left(r - \dfrac{rx}{K} - \dfrac{my}{a+x}\right) + \dfrac{m}{n}\left(\dfrac{nxy}{a+x} - dy\right)$$

$$= 2rx - rK - \dfrac{rx^2}{K} + \dfrac{mKy}{a+x} - \dfrac{dmy}{n}$$

$$= -\dfrac{r}{K}(x - K)^2 + \dfrac{mKy}{a+x} - \dfrac{dmy}{n}$$

$$\leq -\dfrac{r}{K}(x - K)^2 + \dfrac{mKy}{a} - \dfrac{dmy}{n}$$

$$\leq -\left(\dfrac{d}{n} - \dfrac{K}{a}\right) my.$$

Owing to the fact that $n < \dfrac{ad}{K}$, we have $^C D_t^\alpha \mathcal{L}_1(x,y) \leq 0$. In consequence of Lemma 5, E_1 is globally asymptotically stable in the region Ω_1. □

Remark 2. Notice E_1 is locally asymptotically stable if $n < \dfrac{ad}{K} + d$ and is globally asymptotically stable if $n < \dfrac{ad}{K}$. Hence, if the global stability condition is fulfilled, then the local stability is also achieved but not vice versa. This fact reinforces that the global stability condition has a larger attracting region than that of the local stability condition.

Theorem 9. If $(K - \hat{x})(a + \hat{x}) < a^2$, then the first co-existence point $\hat{E} = \left(\hat{x}, \dfrac{(K - \hat{x})(a + \hat{x})r}{mK}\right)$ is globally asymptotically stable in the region Ω_1.

Proof. Let $\hat{E} = (\hat{x}, \hat{y})$ where $\hat{y} = \dfrac{(K - \hat{x})(a + \hat{x})r}{mK}$ and φ is a positive constant. By considering a Lyapunov function $\mathcal{L}_2(x,y) = \left[x - \hat{x} - \hat{x} \ln \dfrac{x}{\hat{x}}\right] + \varphi \left[y - \hat{y} - \hat{y} \ln \dfrac{y}{\hat{y}}\right]$, and using Lemma 4, we get

$$^C D_t^\alpha \mathcal{L}_2(x,y) \leq \left(\dfrac{x - \hat{x}}{x}\right) {}^C D_t^\alpha x + \varphi \left(\dfrac{y - \hat{y}}{y}\right) {}^C D_t^\alpha y$$

$$= (x - \hat{x})\left(r - \dfrac{rx}{K} - \dfrac{my}{a+x}\right) + \varphi(y - \hat{y})\left(\dfrac{nx}{a+x} - d\right)$$

$$= (x - \hat{x})\left(\dfrac{r\hat{x}}{K} + \dfrac{m\hat{y}}{a+\hat{x}} - \dfrac{rx}{K} - \dfrac{my}{a+x}\right) + \varphi(y - \hat{y})\left(\dfrac{nx}{a+x} - \dfrac{n\hat{x}}{a+\hat{x}}\right)$$

$$= -(x - \hat{x})\left((x - \hat{x})\dfrac{r}{K} + \left(\dfrac{y}{a+x} - \dfrac{\hat{y}}{a+\hat{x}}\right)m\right) + n\varphi(y - \hat{y})\left(\dfrac{x}{a+x} - \dfrac{\hat{x}}{a+\hat{x}}\right)$$

$$= -(x - \hat{x})^2 \dfrac{r}{K} - (x - \hat{x})\left(\dfrac{(y - \hat{y})(a + \hat{x}) + (\hat{x} - x)\hat{y}}{(a+x)(a+\hat{x})}\right) m + \left(\dfrac{(x - \hat{x})(y - \hat{y})}{(a+x)(a+\hat{x})}\right) an\varphi$$

$$= -(x - \hat{x})^2 \dfrac{r}{K} - \left(\dfrac{(x - \hat{x})(y - \hat{y})(a + \hat{x})}{(a+x)(a+\hat{x})}\right) m + \left(\dfrac{(x - \hat{x})^2 \hat{y}}{(a+x)(a+\hat{x})}\right) m +$$

$$\left(\dfrac{(x - \hat{x})(y - \hat{y})}{(a+x)(a+\hat{x})}\right) an\varphi$$

$$\leq -(x - \hat{x})^2 \left(\dfrac{r}{K} - \dfrac{m\hat{y}}{a^2}\right) + \left(\dfrac{(x - \hat{x})(y - \hat{y})}{(a+x)(a+\hat{x})}\right)(an\varphi - (a + \hat{x})m).$$

By choosing $\varphi = \dfrac{(a+\hat{x})m}{an}$ and substituting the value of $\hat{y}$, we get

$$^C\mathcal{D}_t^\alpha \mathcal{L}_2(x,y) \leq -(x-\hat{x})^2 \left(a^2 - (K-\hat{x})(a+\hat{x})\right) \dfrac{r}{a^2 K}.$$

Therefore, if $(K-\hat{x})(a+\hat{x}) < a^2$, then $^C\mathcal{D}_t^\alpha \mathcal{L}_2(x,y) \leq 0$. Using Lemma 5, we conclude that $\hat{E}$ is globally asymptotically stable in the region Ω_1. □

Based on Theorem 2, $x(t)$ and $y(t)$ are bounded. Let Ψ be the upper bound of $y(t)$ such that $0 < T \leq y(t) \leq \Psi$. The global stability of E^* is stated in the following theorem.

Theorem 10. *If $y^* < \min\left\{\dfrac{a^2 r}{mK}, \dfrac{cT}{\Psi - T} + T\right\}$, then the second co-existence point $E^* = (x^*, y^*)$ is globally asymptotically stable in the region Ω_2.*

Proof. We consider a positive Lyapunov function $\mathcal{L}_3(E^*) = \left[x - x^* - x^* \ln \dfrac{x}{x^*}\right] + \varphi^* \left[y - y^* - y^* \ln \dfrac{y}{y^*}\right]$. According to Lemma 4, the fractional derivative of $\mathcal{L}_3(E^*)$ satisfies

$$^C\mathcal{D}_t^\alpha \mathcal{L}_2(x,y) \leq \left(\dfrac{x-x^*}{x}\right) {}^C\mathcal{D}_t^\alpha x + \varphi^* \left(\dfrac{y-y^*}{y}\right) {}^C\mathcal{D}_t^\alpha y$$

$$= (x-x^*)\left(r - \dfrac{rx}{K} - \dfrac{my}{a+x}\right) + \varphi^*(y-y^*)\left(\dfrac{nx}{a+x} - d - \dfrac{h(y-T)}{(c+y-T)y}\right)$$

$$= (x-x^*)\left(\dfrac{rx^*}{K} + \dfrac{my^*}{a+x^*} - \dfrac{rx}{K} - \dfrac{my}{a+x}\right)$$
$$+ \varphi^*(y-y^*)\left(\dfrac{nx}{a+x} - \dfrac{nx^*}{a+x^*} + \dfrac{h(y^*-T)}{(c+y^*-T)y^*} - \dfrac{h(y-T)}{(c+y-T)y}\right)$$

$$= -(x-x^*)^2 \dfrac{r}{K} - (x-x^*)\left(\dfrac{(y-y^*)(a+x^*)m - (x-x^*)my^*}{(a+x^*)(a+x)}\right)$$
$$+ \varphi^*(y-y^*)\left(\dfrac{(x-x^*)an}{(a+x^*)(a+x)} - \dfrac{(y-y^*)chT - (y-y^*)(y^*-T)(y-T)h}{(c+y^*-T)(c+y-T)y^*y}\right)$$

$$= -(x-x^*)^2 \dfrac{r}{K} - \dfrac{(x-x^*)(y-y^*)(a+x^*)m}{(a+x^*)(a+x)} + \dfrac{(x-x^*)^2 my^*}{(a+x^*)(a+x)}$$
$$+ \dfrac{(x-x^*)(y-y^*)an\varphi^*}{(a+x^*)(a+x)} - \dfrac{(cT - (y^*-T)(y-T))(y-y^*)^2 h\varphi^*}{(c+y^*-T)(c+y-T)y^*y}$$

$$\leq -(x-x^*)^2 \dfrac{r}{K} - \dfrac{(x-x^*)(y-y^*)(a+x^*)m}{(a+x^*)(a+x)} + (x-x^*)^2 \dfrac{my^*}{a^2}$$
$$+ \dfrac{(x-x^*)(y-y^*)an\varphi^*}{(a+x^*)(a+x)} - \dfrac{(cT - (y^*-T)(y-T))(y-y^*)^2 h\varphi^*}{(c+y^*-T)(c+y-T)y^*y}$$

By taking $\varphi^* = \dfrac{(a+x^*)m}{an}$ and remembering that $y(t) < \Psi$, we obtain

$$^C\mathcal{D}_t^\alpha \mathcal{L}_2(E^*) \leq -(x-x^*)^2\left(\dfrac{r}{K} - \dfrac{my^*}{a^2}\right) - \dfrac{(cT - (y^* - T)(\Psi - T))(y-y^*)^2(a+x^*)m a}{(c+y^*-T)(c+y-T)any^*y}.$$

It is easily confirmed that, if $y^* < \min\left\{\dfrac{a^2 r}{mK}, \dfrac{cT}{\Psi - T} + T\right\}$, then $^C\mathcal{D}_t^\alpha \mathcal{L}_2(x,y) \leq 0$. Based on Lemma 5, $\hat{E}$ is globally asymptotically stable in the region Ω_2. □

2.7. The Existence of Hopf Bifurcation

The Hopf bifurcation is a local phenomenon when a stable equilibrium point loses its stability and all nearby solutions converge to a periodic solution namely limit-cycle if a parameter is varied [54,60,61]. It is shown that many fractional-order models involving the Caputo operator undergo a Hopf bifurcation which is driven by the order of the derivative (see [2,17,34,53,62]). The difference between the Hopf bifurcation in the integer-order model and that in the fractional-order model lies on the property of the limit-cycle. In the integer-order model, the limit-cycle is a periodic orbit which does not exist in the fractional-order model [63]. In the fractional-order model, the limit-cycle is not a periodic solution, but all nearby solutions converge to a limit-cycle [56,62].

Adapted from Theorem 3 in [62], for two dimensional Caputo fractional-order system, a Hopf bifurcation occurs when the eigenvalues $\lambda_{1,2}$ of the Jacobian matrix evaluated at the equilibrium point satisfy the following conditions:

(i) $\lambda_{1,2} = \psi \pm \omega i$ where $\psi > 0$;
(ii) $m(\alpha^*) = \alpha^* \pi / 2 - \min_{1 \leq i \leq 2} |\arg(\lambda_i)| = 0$; and
(iii) $\left. \dfrac{dm(\alpha)}{d\alpha} \right|_{\alpha = \alpha^*} \neq 0$.

Now, consider the stability condition in Theorems 6 and 7. For $y < T$, the Jacobian matrix of model (6) evaluated at $\hat{E}$ has a pair of complex eigenvalues if $\Delta > 0$. We can easily confirm that, if $\hat{x} < \dfrac{K-a}{2}$, then the real part of the eigenvalues are positive. We also have that $m(\hat{\alpha}) = 0$ and $\left. \dfrac{dm(\alpha)}{d\alpha} \right|_{\alpha = \hat{\alpha}} \neq 0$. Therefore, $\hat{E}$ undergoes a Hopf bifurcation when α crosses $\hat{\alpha}$. A similar circumstance also occurs when $y \geq T$. When $\xi^2 < 4\theta$, the Jacobian matrix of model (6) has a pair of complex eigenvalues. The real part of the eigenvalues are also positive when $\xi > 0$. We can also check that $m(\alpha^*) = 0$ and $\left. \dfrac{dm(\alpha)}{d\alpha} \right|_{\alpha = \alpha^*} \neq 0$. This means the Hopf bifurcation also occurs when $y \geq T$. Therefore, we have the following theorem.

Theorem 11. (i) Let $\Delta > 0$ and $\hat{x} < \dfrac{K-a}{2}$. The first co-existence point $\hat{E}$ undergoes a Hopf bifurcation when α passes through $\hat{\alpha}$ in the region Ω_1.
(ii) Let $\xi^2 < 4\theta$ and $\xi > 0$. The second co-existence point E^* undergoes a Hopf bifurcation when α passes through α^* in the region Ω_2.

3. The Atangana–Baleanu Fractional-Order Rosenzweig–MacArthur Model with THP in Predator

3.1. Model Formulation

In this section, we consider a fractional-order Rosenzweig–MacArthur model with THP in predator involving the Atangana–Baleanu operator. Specifically, we consider the Atangana–Baleanu operator in Caputo sense of order-α which is defined as follows.

Definition 3. [40] Suppose $0 < \alpha \leq 1$. The Atangana–Baleanu fractional integral and derivative in Caputo sense of order-α (ABC) is defined by

$$^{ABC}\mathcal{I}^\alpha_t f(t) = \dfrac{1-\alpha}{B(\alpha)} f(t) + \dfrac{\alpha}{\Gamma(\alpha) B(\alpha)} \int_0^t (t-s)^{\alpha-1} f(s) ds,$$

$$^{ABC}\mathcal{D}^\alpha_t f(t) = \dfrac{B(\alpha)}{1-\alpha} \int_0^t E_\alpha \left[-\dfrac{\alpha}{1-\alpha}(t-s)^\alpha \right] f'(s) ds,$$

where $t \geq 0$, $f \in C^n([0,+\infty), \mathbb{R})$, $E_\alpha(t) = \sum_{k=0}^{\infty} \dfrac{t^k}{\Gamma(\alpha k + 1)}$ is the Mittag-Leffler function and $B(\alpha)$ is a normalization function with $B(0) = B(1) = 1$. In this paper, we define $B(\alpha) = 1 - \alpha + \dfrac{\alpha}{\Gamma(\alpha)}$.

By applying Definition 3 to model (4), we get the following fractional-order model with ABC operator

$$^{ABC}\mathcal{D}_t^\alpha x = rx\left(1 - \frac{x}{K}\right) - \frac{mxy}{a+x} \equiv G_1,$$
$$^{ABC}\mathcal{D}_t^\alpha y = \frac{nxy}{a+x} - dy - H(y) \equiv G_2. \tag{16}$$

Due to the lack of analytical theory, model (16) is investigated numerically. However, we first show the existence and uniqueness of the solution of the model (16).

3.2. Existence and Uniqueness

We start this work by representing the Lipschitz condition of the kernels of model (16). Since the harvesting is performed by obeying threshold harvesting policy, we give the proof into twc cases i.e., $y < T$ and $y \geq T$.

We start for $y \geq T$. Let $x, \tilde{x}, y,$ and $\tilde{y}$ be functions satisfying $\|x\| \leq a_1, \|\tilde{x}\| \leq a_2, |y| \leq b_1,$ and $\|\tilde{y}\| \leq b_2$. Suppose that

$$g_1 = r + (a_1 + a_2)\frac{r}{K} + \frac{my}{a},$$
$$g_2 = n + d + \frac{h}{c}.$$

Therefore, we get

$$\begin{aligned}
\|G_1(x) - G_1(\tilde{x})\| &= \left\|\left(rx\left(1 - \frac{x}{K}\right) - \frac{mxy}{a+x}\right) - \left(r\tilde{x}\left(1 - \frac{\tilde{x}}{K}\right) - \frac{m\tilde{x}y}{a+\tilde{x}}\right)\right\| \\
&= \left\|rx - \frac{rx^2}{K} - \frac{mxy}{a+x} - r\tilde{x} + \frac{r\tilde{x}^2}{K} + \frac{m\tilde{x}y}{a+\tilde{x}}\right\| \\
&= \left\|r(x - \tilde{x}) - \frac{r}{K}(x^2 - \tilde{x}^2) - my\left(\frac{x}{a+x} - \frac{\tilde{x}}{a+\tilde{x}}\right)\right\| \\
&= \left\|r(x - \tilde{x}) - \frac{r}{K}(x + \tilde{x})(x - \tilde{x}) - \frac{amy(x - \tilde{x})}{(a+x)(a+\tilde{x})}\right\| \\
&\leq r\|x - \tilde{x}\| + (a_1 + a_2)\frac{r}{K}\|x - \tilde{x}\| + \frac{my}{a}\|x - \tilde{x}\| \\
&= \left(r + (a_1 + a_2)\frac{r}{K} + \frac{my}{a}\right)\|x - \tilde{x}\| \\
&= g_1 \|x - \tilde{x}\|,
\end{aligned} \tag{17}$$

and

$$\begin{aligned}
\|G_2(y) - G_2(\tilde{y})\| &= \left\|\left(\frac{nxy}{a+x} - dy - \frac{h(y-T)}{c+(y-T)}\right) - \left(\frac{nx\tilde{y}}{a+x} - d\tilde{y} - \frac{h(\tilde{y}-T)}{c+(\tilde{y}-T)}\right)\right\| \\
&= \left\|\frac{nxy}{a+x} - dy - \frac{h(y-T)}{c+(y-T)} - \frac{nx\tilde{y}}{a+x} + d\tilde{y} + \frac{h(\tilde{y}-T)}{c+(\tilde{y}-T)}\right\| \\
&= \left\|\frac{nx(y-\tilde{y})}{a+x} - d(y-\tilde{y}) - \frac{ch(y-\tilde{y})}{(c+y-T)(c+\tilde{y}-T)}\right\| \\
&\leq n\|y - \tilde{y}\| + d\|y - \tilde{y}\| + \frac{h}{c}\|y - \tilde{y}\| \\
&= \left(n + d + \frac{h}{c}\right)\|y - \tilde{y}\| \\
&= g_2 \|y - \tilde{y}\|.
\end{aligned} \tag{18}$$

When $y < T$, by utilizing the similar manner, we achieve $\|G_2(y) - G_2(\bar{y})\| \leq g_3 \|y - \bar{y}\|$ where $g_3 = n + d$. Accordingly, the kernel of (16) satisfies the Lipschitz condition. Furthermore, if $0 \leq g_1 < 1$ and $0 \leq g_3 < g_2 < 1$, then G_1 and G_2 are contracted.

Now, by employing the fixed-point theorem, the solution of model (16) is investigated. By utilizing the Atangana–Baleanu fractional integral operator, model (16) is transformed into the following Volterra-type integral equation.

$$\begin{aligned} x(t) - x(0) &= \frac{1-\alpha}{B(\alpha)} G_1(t,x) + \frac{\alpha}{B(\alpha)\Gamma(\alpha)} \int_0^t (t-s)^{\alpha-1} G_1(s,x) ds, \\ y(t) - y(0) &= \frac{1-\alpha}{B(\alpha)} G_2(t,y) + \frac{\alpha}{B(\alpha)\Gamma(\alpha)} \int_0^t (t-s)^{\alpha-1} G_2(s,y) ds. \end{aligned} \tag{19}$$

In a recursive form, Equation (19) is written as

$$\begin{aligned} x_n(t) &= \frac{1-\alpha}{B(\alpha)} G_1(t, x_{n-1}) + \frac{\alpha}{B(\alpha)\Gamma(\alpha)} \int_0^t (t-s)^{\alpha-1} G_1(s, x_{n-1}) ds, \\ y_n(t) &= \frac{1-\alpha}{B(\alpha)} G_2(t, y_{n-1}) + \frac{\alpha}{B(\alpha)\Gamma(\alpha)} \int_0^t (t-s)^{\alpha-1} G_2(s, y_{n-1}) ds. \end{aligned} \tag{20}$$

The associated initial conditions along with Equation (20) are $x_0(t) = x(0)$ and $y_0(t) = y(0)$. By considering Equation (20), we have the difference expression of successive terms as follows.

$$\begin{aligned} \Phi_{1,n}(t) &= x_n(t) - x_{n-1}(t) \\ &= \frac{1-\alpha}{B(\alpha)} \left(G_1(t, x_{n-1}) - G_1(t, x_{n-2}) \right) + \frac{\alpha}{B(\alpha)\Gamma(\alpha)} \int_0^t (t-s)^{\alpha-1} \left(G_1(t, x_{n-1}) - G_1(t, x_{n-2}) \right) ds, \\ \Phi_{2,n}(t) &= y_n(t) - y_{n-1}(t) \\ &= \frac{1-\alpha}{B(\alpha)} \left(G_2(t, y_{n-1}) - G_2(t, y_{n-2}) \right) + \frac{\alpha}{B(\alpha)\Gamma(\alpha)} \int_0^t (t-s)^{\alpha-1} \left(G_2(t, y_{n-1}) - G_2(t, y_{n-2}) \right) ds. \end{aligned} \tag{21}$$

According to Equation (21), we get $x_n(t) = \sum_{i=1}^n \Phi_{1,i}(t)$ and $y_n(t) = \sum_{i=1}^n \Phi_{2,i}(t)$. By applying Equations (17), (18) and (21), we have that

$$\begin{aligned} \|\Phi_{1,n}(t)\| &\leq \frac{1-\alpha}{B(\alpha)} g_1 \|\Phi_{1,n-1}\| + \frac{\alpha}{B(\alpha)\Gamma(\alpha)} g_1 \int_0^t \|\Phi_{1,n-1}(s)\| (t-s)^{\alpha-1} ds, \\ \|\Phi_{2,n}(t)\| &\leq \frac{1-\alpha}{B(\alpha)} g_2 \|\Phi_{2,n-1}\| + \frac{\alpha}{B(\alpha)\Gamma(\alpha)} g_2 \int_0^t \|\Phi_{2,n-1}(s)\| (t-s)^{\alpha-1} ds. \end{aligned} \tag{22}$$

Therefore, by using (22), the existence and uniqueness of model (16) is presented as follows.

Theorem 12. *Model (16) has a unique solution if we can find t_0 such that*

$$\frac{(1-\alpha)g_i}{B(\alpha)} + \frac{t_0^\alpha g_i}{B(\alpha)\Gamma(\alpha)} < 1, i = 1, 2, 3. \tag{23}$$

Proof. Let $x(t)$ and $y(t)$ be bounded functions, and hence the Lipschitz condition is satisfied. Thus, according to Equation (22), we obtain the following inequalities.

$$\begin{aligned} \|\Phi_{1,n}(t)\| &\leq \|x_0\| \left(\frac{(1-\alpha)g_1}{B(\alpha)} + \frac{t^\alpha g_1}{B(\alpha)\Gamma(\alpha)} \right)^n, \\ \|\Phi_{2,n}(t)\| &\leq \|y_0\| \left(\frac{(1-\alpha)g_2}{B(\alpha)} + \frac{t^\alpha g_2}{B(\alpha)\Gamma(\alpha)} \right)^n. \end{aligned} \tag{24}$$

Therefore, the continuity and existence of solution are proved since $\|\Phi_{1,n}(t)\| \to 0$ and $\|\Phi_{2,n}(t)\| \to 0$ as $n \to \infty$ and $t = t_0$. Now, suppose that

$$\begin{aligned} x(t) - x(0) &= x_n(t) - Y_{1,n}(t), \\ y(t) - y(0) &= y_n(t) - Y_{2,n}(t). \end{aligned} \quad (25)$$

We confirm that

$$\|Y_{1,n}(t)\| \leq \left(\frac{1-\alpha}{B(\alpha)} + \frac{t^\alpha}{B(\alpha)\Gamma(\alpha)}\right)^{n+1} g_1^{n+1} \quad (26)$$

It is clear that $\|Y_{1,n}(t)\| \to 0$ when $n \to \infty$. By using a similar manner, we acquire $\|Y_{2,n}(t)\| \to 0, n \to \infty$. Finally, the uniqueness of solution for the model is proven. Suppose that there exists different set of solutions denote by $\tilde{x}(t)$ and $\tilde{y}(t)$; then, we have

$$x(t) - \tilde{x}(t) = \frac{1-\alpha}{B(\alpha)}(G_1(t,x) - G_1(t,\tilde{x})) + \frac{\alpha}{B(\alpha)\Gamma(\alpha)} \int_0^t (G_1(s,x) - G_1(s,\tilde{x}))(t-s)^{\alpha-1} ds. \quad (27)$$

Taking the norm for both sides and using a simplification as in (22) and (24), we obtain

$$\|x(t) - \tilde{x}(t)\| \left(1 - \frac{(1-\alpha)g_1}{B(\alpha)} - \frac{t^\alpha g_1}{B(\alpha)\Gamma(\alpha)}\right) \leq 0. \quad (28)$$

For $t = t_0$, we have (23), thus $\|x(t) - \tilde{x}(t)\| = 0$ and hence $x(t) = \tilde{x}(t)$. Applying the same algebraic procedures, we can show that $y(t) = \tilde{y}(t)$. Therefore, the solution is unique. □

4. Numerical Simulations

In this section, the numerical simulations of Caputo model (6) and ABC model (16) are performed to illustrate the previous theoretical results. In the literature, there exist many numerical methods to solve a system of fractional differential equations, such as the Monte Carlo method [64], the Grünwald–Letnikov method [65,66] and the predictor–corrector method [67–69]. We apply the predictor–corrector scheme proposed by Diethelm et al. [67] to solve the Caputo fractional-order model and the predictor–corrector scheme proposed by Baleanu et al. [69] to solve the Atangana–Baleanu in Caputo sense model (ABC). Due to the limitation of field data, we use hypothetical parameter values that correspond to the theoretical results. The initial parameter values are given as follows.

$$r = 0.5, K = 1, m = 0.3, a = 0.2, d = 0.1, h = 0.1, T = 0.9, c = 0.1, \text{ and } \alpha = 0.9. \quad (29)$$

In Figure 1, we plot a bifurcation diagram by varying the conversion rate of consumed prey into predator birth n in interval $[0.08, 0.2]$. We notice that the parameter values (29) and the interval $0.08 \leq n \leq 0.2$ lead to the non-existence of equilibrium point in Ω_2. Therefore, the first numerical simulations are focused on the dynamics in Ω_1.

For $0.08 \leq n < n_1^* = 0.12$, Theorem 5 states that the predator extinction point $E_1 = (1,0)$ is the only equilibrium point which is asymptotically stable. To see this behavior, we take $n = 0.1$ and plot the phase portrait and the time series as in Figure 2. It is seen that all solutions with initial values in both Ω_1 and Ω_2 are convergent to E_1. When the initial value is in Ω_2, then the solution is oscillating when it crosses the threshold harvesting level and eventually converges to E_1.

When n passes through n_1^*, the equilibrium point $E_1 = (1,0)$ undergoes a forward bifurcation. In this case, there appear two equilibrium points, namely the unstable E_1 and an asymptotically stable $\hat{E}$. Figure 1 shows that $\hat{E}$ is asymptotically stable if $0.12 < n \lesssim n_2^* = 0.1557$. In Figure 3, we show the phase portrait and time series for the case of $n = 0.14$. We see that $E_0 = (0,0)$ and $E_1 = (1,0)$ are saddle points, while $\hat{E} = (0.5, 0.5833)$ is asymptotically stable. This circumstance corresponds to Theorems 4–6 and 9.

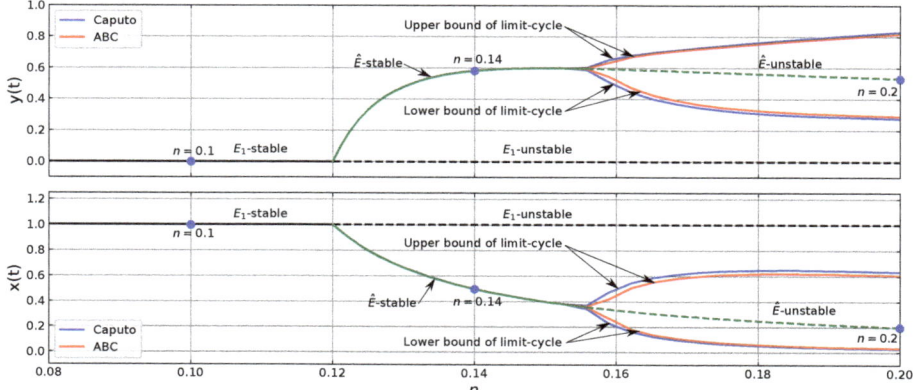

Figure 1. Bifurcation diagram of Caputo model (6) and ABC model (16) driven by the conversion rate of consumed prey into predator birth (n) with constant parameter values (29). There exists two bifurcations namely a forward bifurcation which occurs when n passes through $n_1^* \approx 0.12$, and a Hopf bifurcation which occurs when n passes through $n_2^* \approx 0.1557$.

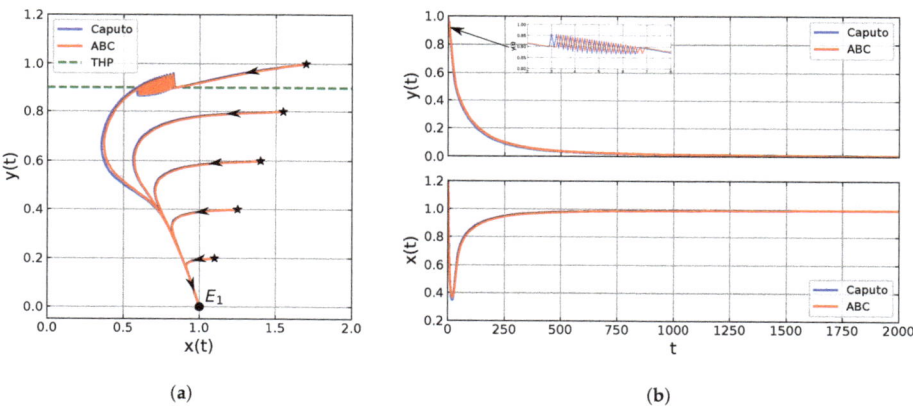

Figure 2. Numerical simulations of Caputo model (6) and ABC model (16) with parameter values (29) and $n = 0.1$: Figure (**a**) phase portrait; and Figure (**b**) time series.

Furthermore, if we increase the value of n such that $n > n_2^*$, then the equilibrium point $\hat{E}$ loses its stability, and all solutions converge to a limit-cycle. This situation confirms the occurrence of a Hopf bifurcation driven by n. For example, if we select $n = 0.2$ then all equilibrium points are unstable and all solutions eventually converge to limit cycle (see Figure 4).

Now, we perform simulation using the same parameter values as in Figure 4, but with a lower threshold value, namely $T = 0.5$. In this case, there is no equilibrium point $\hat{E}$ in Ω_1, and $E^* = (0.5954, 0.5364)$ occurs in Ω_2. According to Theorem 7, E^* is asymptotically stable. Such dynamics can be clearly seen in Figure 5. This simulation shows that by applying the THP when the interior equilibrium point is stable, we can choose a suitable constant of threshold so that the existence of both prey and predator are maintained.

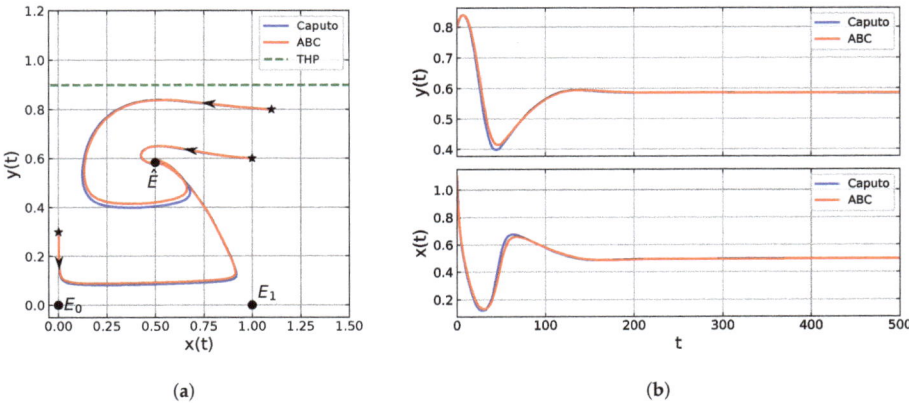

Figure 3. Numerical simulations of Caputo model (6) and ABC model (16) with parameter values (29) and $n = 0.14$: Figure (**a**) phase portrait; and Figure (**b**) time series.

Notice that, in Figures 2–5, we see that both model with Caputo operator (6) and Atangana–Baleanu operator (16) have similar dynamical behavior. The noticeable difference between them is the orbit of solutions and the diameter of the limit-cycle. In Figure 4, the diameter of the limit-cycle of the model with ABC operator looks shorter than that of the Caputo operator, which may give different dynamics when a Hopf bifurcation occurs. To get more detail view, we plot a bifurcation diagram by varying the order of the fractional derivative (α) (see Figure 6). In this simulation, we use parameter values as in Figure 4 and vary the order-α in the interval $[0.6, 0.9]$. It is clearly seen that, besides the diameter of the limit-cycle, the bifurcation points of Caputo model and ABC model are also different. The Caputo model has an earlier bifurcation point than that of the ABC model. To show this situation, we show some numerical simulations using different values of α (see Figure 7). For $\alpha = 0.7$, the equilibrium point $\hat{E}$ of both model are asymptotically stable. For $\alpha = 0.772$, the equilibrium point $\hat{E}$ of the Caputo model loses its stability, while that of the ABC model remains asymptotically stable. For $\alpha = 0.83$, the equilibrium point $\hat{E}$ of both models are unstable.

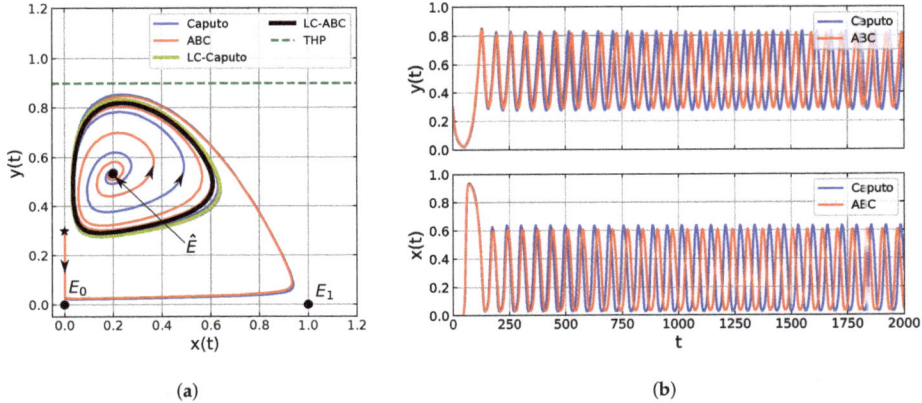

Figure 4. Numerical simulations of Caputo model (6) and ABC model (16) with parameter values (29) and $n = 0.2$: Figure (**a**) phase portrait; and Figure (**b**) time series.

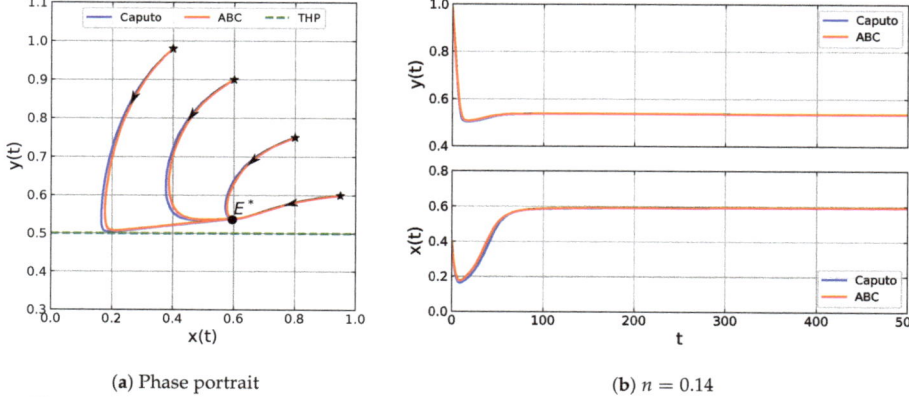

(a) Phase portrait (b) $n = 0.14$

Figure 5. Numerical simulations of Caputo model (6) and ABC model (16) with parameter values (29), $n = 0.2$ and $T = 0.5$.

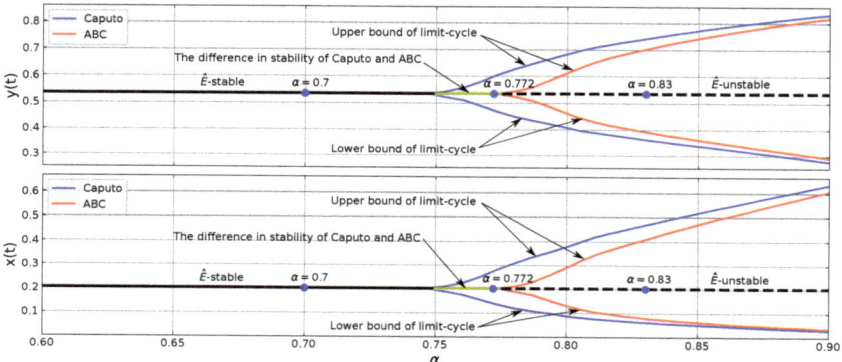

Figure 6. Bifurcation diagram of Caputo model (6) and ABC model (16) driven by the order of the fractional-derivative (α) with constant parameter values (29) and $n = 0.2$. There exists a Hopf bifurcation where the bifurcation points of the Caputo model and ABC model are different.

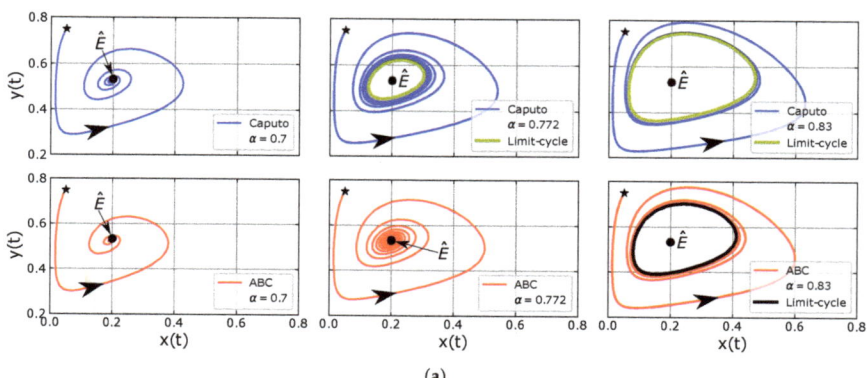

(a)

Figure 7. Cont.

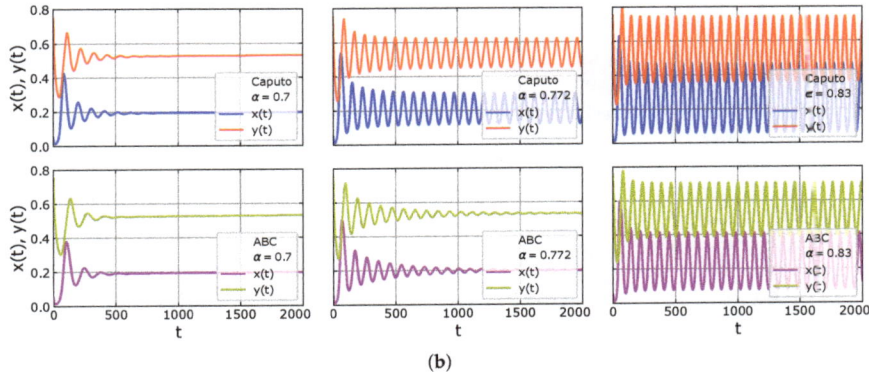

(b)

Figure 7. Numerical simulations of Caputo model (6) and ABC model (16) with parameter values (29), $n = 0.2$, and $\alpha = \{0.7, 0.772, 0.83\}$: Figure (a) phase portrait; and Figure (b) time series.

From the biological point of view, all previous numerical simulations show that the dynamical properties of both Caputo model and ABC model are similar except when the eigenvalues of the Jacobian matrix evaluated at the interior equilibrium point $\hat{E}$ are a pair of complex conjugate with positive real part. There is a biological condition such that the prey and predator densities are eventually periodic for the Caputo model, while for ABC model, the densities of both predator and prey are eventually constant.

5. Conclusions

The dynamics of a Rosenzweig–MacArthur model with continuous threshold harvesting in predator involving the Caputo fractional-order derivative and ABC fractional-order derivative are studied. We prove the existence and uniqueness of solutions of both Caputo and ABC models. Particularly, we completely investigate the dynamics of the Caputo model including the non-negativity, boundedness, local stability, global stability, and the existence of Hopf bifurcation. From the biological meanings, the extinction of both populations never occurs since the origin point (E_0) is a saddle point. Some of the situations that might occur are: (1) the predator goes extinct while the prey still survives, which is indicated by the stability of E_1; (2) both predator and prey co-exist and converge to a constant population density, which happens if the interior point $\hat{E}$ or E^* are asymptotically stable; and (3) both predator and prey co-exist where both population densities change periodically, namely when a Hopf bifurcation occurs. We show numerically that our model may undergo a forward bifurcation or a Hopf bifurcation. The Hopf bifurcation in models with both Caputo operator and ABC operator can be driven by the conversion rate of consumed prey into the predator birth rate or by the order of fractional derivative. Our numerical simulations show that the Hopf bifurcation point of both models are different.

Author Contributions: All authors equally contributed to this article. All authors have read and agreed to the published version of the manuscript.

Funding: This research was funded by the Directorate of Research and Community Service, The Directorate General of Strengthening Research and Development, the Ministry of Research, Technology, and Higher Education (Brawijaya University), Indonesia, via Doctoral Dissertation Research, in accordance with the Research Contract No. 037/SP2H/LT/DRPM/2020, dated 9 March 2020.

Acknowledgments: The authors would like to thank the reviewers for all useful and helpful comments to improve the manuscript.

Conflicts of Interest: All authors report no conflict of interest relevant to this article.

References

1. Rosenzweig, M.L.; MacArthur, R.H. Graphical representation and stability conditions of predator-prey interactions. *Am. Nat.* **1963**, *97*, 209–223. [CrossRef]
2. Moustafa, M.; Mohd, M.H.; Ismail, A.I.; Abdullah, F.A. Stage structure and refuge effects in the dynamical analysis of a fractional order Rosenzweig-MacArthur prey-predator model. *Prog. Fract. Differ. Appl.* **2019**, *5*, 49–64. [CrossRef]
3. Beay, L.K.; Suryanto, A.; Darti, I.; Trisilowati, T. Hopf bifurcation and stability analysis of the Rosenzweig-MacArthur predator-prey model with stage-structure in prey. *Math. Biosci. Eng.* **2020**, *17*, 4080–4097. [CrossRef] [PubMed]
4. González-Olivares, E.; Ramos-Jiliberto, R. Dynamic consequences of prey refuges in a simple model system: More prey, fewer predators and enhanced stability. *Ecol. Model.* **2003**, *166*, 135–146. [CrossRef]
5. Chen, L.; Chen, F.; Chen, L. Qualitative analysis of a predator–prey model with Holling type II functional response incorporating a constant prey refuge. *Nonlinear Anal. Real World Appl.* **2010**, *11*, 246–252. [CrossRef]
6. Almanza-Vasquez, E.; Ortiz-Ortiz, R.D.; Marin-Ramirez, A.M. Bifurcations in the dynamics of Rosenzweig-Macarthur predator-prey model considering saturated refuge for the preys. *Appl. Math. Sci.* **2015**, *9*, 7475–7482. [CrossRef]
7. Das, M.; Maiti, A.; Samanta, G.P. Stability analysis of a prey-predator fractional order model incorporating prey refuge. *Ecol. Genet. Genom.* **2018**, *7-8*, 33–46. [CrossRef]
8. Moustafa, M.; Mohd, M.H.; Ismail, A.I.; Abdullah, F.A. Dynamical analysis of a fractional-order Rosenzweig–MacArthur model incorporating a prey refuge. *Chaos Solitons Fractals* **2018**, *109*, 1–13. [CrossRef]
9. Sarkar, K.; Khajanchi, S. Impact of fear effect on the growth of prey in a predator-prey interaction model. *Ecol. Complex.* **2020**, *42*, 100826. [CrossRef]
10. Zu, J.; Mimura, M. The impact of Allee effect on a predator-prey system with Holling type II functional response. *Appl. Math. Comput.* **2010**, *217*, 3542–3556. [CrossRef]
11. Pal, P.J.; Saha, T. Qualitative analysis of a predator-prey system with double Allee effect in prey. *Chaos Solitons Fractals* **2015**, *73*, 36–63. [CrossRef]
12. Bodine, E.N.; Yust, A.E. Predator–prey dynamics with intraspecific competition and an Allee effect in the predator population. *Lett. Biomath.* **2017**, *4*, 23–38. [CrossRef]
13. Mukherjee, D. Role of fear in predator–prey system with intraspecific competition. *Math. Comput. Simul.* **2020**, *177*, 263–275. [CrossRef]
14. Van Den Bosch, F. Cannibalism in an age-structured predator-prey system. *Bull. Math. Biol.* **1997**, *59*, 551–567. [CrossRef]
15. Mondal, S.; Lahiri, A.; Bairagi, N. Analysis of a fractional order eco-epidemiological model with prey infection and type 2 functional response. *Math. Methods Appl. Sci.* **2017**, *40*, 6776–6789. [CrossRef]
16. Suryanto, A.; Darti, I.; Anam, S. Stability analysis of pest-predator interaction model with infectious disease in prey. *AIP Conf. Proc.* **2018**, *1937*, 020018. [CrossRef]
17. Panigoro, H.S.; Suryanto, A.; Kusumawinahyu, W.M.; Darti, I. Dynamics of a fractional-order predator-prey model with infectious diseases in prey. *Commun. Biomath. Sci.* **2019**, *2*, 105–117. [CrossRef]
18. Kumar, T.; And, K.; Chakraborty, K. Effort dynamics in a prey-predator model with harvesting. *Int. J. Inf. Syst. Sci.* **2000**, *6*, 318–332.
19. Javidi, M.; Nyamoradi, N. Dynamic analysis of a fractional order prey-predator interaction with harvesting. *Appl. Math. Model.* **2013**, *37*, 8946–8956. [CrossRef]
20. Zhu, C.; Kong, L. Bifurcations analysis of Leslie-Gower predator-prey models with nonlinear predator-harvesting. *Discret. Contin. Dyn. Syst.* **2017**, *10*, 1187–1206. [CrossRef]
21. Suryanto, A.; Darti, I. Dynamics of Leslie-Gower pest-predator model with disease in pest including pest-harvesting and optimal implementation of pesticide. *Int. J. Math. Math. Sci.* **2019**, *2019*, 1–9. [CrossRef]
22. Ang, T.K.; Safuan, H.M. Harvesting in a toxicated intraguild predator–prey fishery model with variable carrying capacity. *Chaos Solitons Fractals* **2019**, *126*, 158–168. [CrossRef]
23. Leard, B.; Rebaza, J. Analysis of predator-prey models with continuous threshold harvesting. *Appl. Math. Comput.* **2011**, *217*, 5265–5278. [CrossRef]
24. Bohn, J.; Rebaza, J.; Speer, K. Continuous threshold prey harvesting in predator-prey models. *Int. J. Math. Comput. Sci.* **2011**, *5*, 964–971.

25. Rebaza, J. Dynamics of prey threshold harvesting and refuge. *J. Comput. Appl. Math.* **2012**, *236*, 1743–1752. [CrossRef]
26. Lv, Y.; Yuan, R.; Pei, Y. Dynamics in two nonsmooth predator–prey models with threshold harvesting. *Nonlinear Dyn.* **2013**, *74*, 107–132. [CrossRef]
27. Wu, D.; Zhao, H.; Yuan, Y. Complex dynamics of a diffusive predator–prey model with strong Allee effect and threshold harvesting. *J. Math. Anal. Appl.* **2019**, *469*, 982–1014. [CrossRef]
28. Toaha, S. The effect of harvesting with threshold on the dynamics of prey predator model. *J. Phys. Conf. Ser.* **2019**, *1341*. [CrossRef]
29. Panigoro, H.S.; Suryanto, A.; Kusumawinahyu, W.M.; Darti, I. Continuous threshold harvesting in a gause-type predator-prey model with fractional-order. *AIP Conf. Proc.* **2020**, *2264*, 040001. [CrossRef]
30. Shepard, B.M.; Carner, G.R.; Barrion, A.T.; Ooi, P.A.C.; Van den Berg, H. *Insects and Their Natural Enemies Associated with Vegetables and Soybean in Southeast Asia*; Clemson Univ. Coastal Research &: Orangeburg, SC, USA, 1999.
31. Li, H.L.; Zhang, L.; Hu, C.; Jiang, Y.L.; Teng, Z. Dynamical analysis of a fractional-order predator-prey model incorporating a prey refuge. *J. Appl. Math. Comput.* **2017**, *54*, 435–449. [CrossRef]
32. Das, S.; Gupta, P.K. A mathematical model on fractional Lotka-Volterra equations. *J. Theor. Biol.* **2011**, *277*, 1–6. [CrossRef]
33. Panja, P. Dynamics of a fractional order predator-prey model with intraguild predation. *Int. J. Model. Simul.* **2019**, *39*, 256–268. [CrossRef]
34. Suryanto, A.; Darti, I.; Panigoro, H.S.; Kilicman, A. A fractional-order predator–prey model with ratio-dependent functional response and linear harvesting. *Mathematics* **2019**, *7*, 1100. [CrossRef]
35. Alidousti, J.; Ghafari, E. Dynamic behavior of a fractional order prey-predator model with group defense. *Chaos Solitons Fractals* **2020**, *134*, 109688. [CrossRef]
36. Ghanbari, B.; Djilali, S. Mathematical analysis of a fractional-order predator-prey model with prey social behavior and infection developed in predator population. *Chaos Solitons Fractals* **2020**, *138*, 109960. [CrossRef]
37. Xie, Y.; Wang, Z.; Meng, B.; Huang, X. Dynamical analysis for a fractional-order prey–predator model with Holling III type functional response and discontinuous harvest. *Appl. Math. Lett.* **2020**, *105*, 106342. [CrossRef]
38. Sekerci, Y. Climate change effects on fractional order prey-predator model. *Chaos Solitons Fractals* **2020**, *134*, 109690. [CrossRef]
39. Caputo, M. Linear models of dissipation whose Q is almost fFrequency independent–II. *Geophys. J. Int.* **1967**, *13*, 529–539. [CrossRef]
40. Atangana, A.; Baleanu, D. New fractional derivatives with nonlocal and non-singular kernel: Theory and application to heat transfer model. *Therm. Sci.* **2016**, *20*, 763–769. [CrossRef]
41. Atangana, A.; Koca, I. Chaos in a simple nonlinear system with Atangana–Baleanu derivatives with fractional order. *Chaos Solitons Fractals* **2016**, *89*, 447–454. [CrossRef]
42. Tajadodi, H. A Numerical approach of fractional advection-diffusion equation with Atangana–Baleanu derivative. *Chaos Solitons Fractals* **2020**, *130*, 109527. [CrossRef]
43. Ghanbari, B.; Kumar, D. Numerical solution of predator-prey model with Beddington-DeAngelis functional response and fractional derivatives with Mittag-Leffler kernel. *Chaos Interdiscip. J. Nonlinear Sci.* **2019**, *29*, 063103. [CrossRef] [PubMed]
44. Morales-Delgado, V.F.; Gómez-Aguilar, J.F.; Saad, K.; Escobar Jiménez, R.F. Application of the Caputo-Fabrizio and Atangana-Baleanu fractional derivatives to mathematical model of cancer chemotherapy effect. *Math. Methods Appl. Sci.* **2019**, *42*, 1167–1193. [CrossRef]
45. Shah, S.A.A.; Khan, M.A.; Farooq, M.; Ullah, S.; Alzahrani, E.O. A fractional order model for Hepatitis B virus with treatment via Atangana–Baleanu derivative. *Phys. A Stat. Mech. Its Appl.* **2020**, *538*, 122636. [CrossRef]
46. Bourafa, S.; Abdelouahab, M.S.; Moussaoui, A. On some extended Routh–Hurwitz conditions for fractional-order autonomous systems of order $\alpha \in (0, 2)$ and their applications to some population dynamic models. *Chaos Solitons Fractals* **2020**, *133*. [CrossRef]
47. Ahmed, E.; El-Sayed, A.; El-Saka, H.A. On some Routh–Hurwitz conditions for fractional order differential equations and their applications in Lorenz, Rössler, Chua and Chen systems. *Phys. Lett. A* **2006**, *358*, 1–4. [CrossRef]

48. Petras, I. *Fractional-Order Nonlinear Systems: Modeling, Analysis and Simulation*; Springer: London, UK; Beijing, China, 2011.
49. Diethelm, K. *The Analysis of Fractional Differential Equations: An Application-Oriented Exposition Using Differential Operators of Caputo Type*; Springer: Braunschweig, Germany, 2010.
50. Podlubny, I. *Fractional Differential Equations: An Introduction to Fractional Derivatives, Fractional Differential Equations, to Methods of Their Solution and Some of Their Applications*; Academic Press: San Diego CA, USA, 1999.
51. Vargas-De-León, C. Volterra-type Lyapunov functions for fractional-order epidemic systems. *Commun. Nonlinear Sci. Numer. Simul.* **2015**, *24*, 75–85. [CrossRef]
52. Huo, J.; Zhao, H.; Zhu, L. The effect of vaccines on backward bifurcation in a fractional order HIV model. *Nonlinear Anal. Real World Appl.* **2015**, *26*, 289–305. [CrossRef]
53. Abdelouahab, M.S.; Hamri, N.E.; Wang, J. Hopf bifurcation and caos in fractional-order modified hybrid optical system. *Nonlinear Dyn.* **2012**, *69*, 275–284. [CrossRef]
54. Deshpande, A.S.; Daftardar-Gejji, V.; Sukale, Y.V. On Hopf bifurcation in fractional dynamical systems. *Chaos Solitons Fractals* **2017**, *98*, 189–198. [CrossRef]
55. Diethelm, K. A fractional calculus based model for the simulation of an outbreak of dengue fever. *Nonlinear Dyn.* **2013**, *71*, 613–619. [CrossRef]
56. Moustafa, M.; Mohd, M.H.; Ismail, A.I.; Abdullah, F.A. Dynamical analysis of a fractional-order eco-epidemiological model with disease in prey population. *Adv. Differ. Equ.* **2020**, *2020*, 48. [CrossRef]
57. Li, Y.; Chen, Y.; Podlubny, I. Stability of fractional-order nonlinear dynamic systems: Lyapunov direct method and generalized Mittag–Leffler stability. *Comput. Math. Appl.* **2010**, *59*, 1810–1821. [CrossRef]
58. Odibat, Z.M.; Shawagfeh, N.T. Generalized Taylor's formula. *Appl. Math. Comput.* **2007**, *186*, 286–293. [CrossRef]
59. Matignon, D. Stability results for fractional differential equations with applications to control processing. *Comput. Eng. Syst. Appl.* **1996**, *2*, 963–968.
60. Kuznetsov, Y.A. *Elements of Applied Bifurcation Theory*, 3rd ed.; Springer: New York, NY, USA, 2004.
61. Baisad, K.; Moonchai, S. Analysis of stability and Hopf bifurcation in a fractional Gauss-type predator–prey model with Allee effect and Holling type-III functional response. *Adv. Differ. Equ.* **2018**, *2018*. [CrossRef]
62. Li, X.; Wu, R. Hopf bifurcation analysis of a new commensurate fractional-order hyperchaotic system. *Nonlinear Dyn.* **2014**, *78*, 279–288. [CrossRef]
63. Tavazoei, M.S.; Haeri, M. A proof for non existence of periodic solutions in time invariant fractional order systems. *Automatica* **2009**, *45*, 1886–1890. [CrossRef]
64. Fulger, D.; Scalas, E.; Germano, G. Monte Carlo simulation of uncoupled continuous-time random walks yielding a stochastic solution of the space-time fractional diffusion equation. *Phys. Rev. E* **2008**, *77*, 021122. [CrossRef]
65. Scherer, R.; Kalla, S.L.; Tang, Y.; Huang, J. The Grünwald–Letnikov method for fractional differential equations. *Comput. Math. Appl.* **2011**, *62*, 902–917. [CrossRef]
66. Suryanto, A.; Darti, I. Stability analysis and nonstandard Grünwald-Letnikov scheme for a fractional order predator-prey model with ratio-dependent functional response. *AIP Conf. Proc.* **2017**, *1913*, 020011. [CrossRef]
67. Diethelm, K.; Ford, N.J.; Freed, A.D. A predictor-corrector approach for the numerical solution of fractional differential equations. *Nonlinear Dyn.* **2002**, *29*, 3–22. [CrossRef]
68. Wang, Z. A numerical method for delayed fractional-order differential equations. *J. Appl. Math.* **2013**, *2013*, 256071. [CrossRef]
69. Baleanu, D.; Jajarmi, A.; Hajipour, M. On the nonlinear dynamical systems within the generalized fractional derivatives with Mittag–Leffler kernel. *Nonlinear Dyn.* **2018**, *94*, 397–414. [CrossRef]

© 2020 by the authors. Licensee MDPI, Basel, Switzerland. This article is an open access article distributed under the terms and conditions of the Creative Commons Attribution (CC BY) license (http://creativecommons.org/licenses/by/4.0/).

Article

Inverse Problem for a Mixed Type Integro-Differential Equation with Fractional Order Caputo Operators and Spectral Parameters

Tursun K. Yuldashev [1,*] and Erkinjon T. Karimov [2]

[1] Uzbek-Israel Joint Faculty of High Technology and Engineering Mathematics, National University of Uzbekistan, Tashkent 100174, Uzbekistan
[2] V. I. Romanovskiy Institute of Mathematics, Uzbekistan Academy of Sciences, Tashkent 100174, Uzbekistan; erkinjon@gmail.com
* Correspondence: t.yuldashev@nuu.uz or tursun.k.yuldashev@gmail.com; Tel.: +998-99-519-59-31

Received: 25 September 2020; Accepted: 16 October 2020; Published: 20 October 2020

Abstract: The questions of the one-value solvability of an inverse boundary value problem for a mixed type integro-differential equation with Caputo operators of different fractional orders and spectral parameters are considered. The mixed type integro-differential equation with respect to the main unknown function is an inhomogeneous partial integro-differential equation of fractional order in both positive and negative parts of the multidimensional rectangular domain under consideration. This mixed type of equation, with respect to redefinition functions, is a nonlinear Fredholm type integral equation. The fractional Caputo operators' orders are smaller in the positive part of the domain than the orders of Caputo operators in the negative part of the domain under consideration. Using the method of Fourier series, two systems of countable systems of ordinary fractional integro-differential equations with degenerate kernels and different orders of integro-differentation are obtained. Furthermore, a method of degenerate kernels is used. In order to determine arbitrary integration constants, a linear system of functional algebraic equations is obtained. From the solvability condition of this system are calculated the regular and irregular values of the spectral parameters. The solution of the inverse problem under consideration is obtained in the form of Fourier series. The unique solvability of the problem for regular values of spectral parameters is proved. During the proof of the convergence of the Fourier series, certain properties of the Mittag–Leffler function of two variables, the Cauchy–Schwarz inequality and Bessel inequality, are used. We also studied the continuous dependence of the solution of the problem on small parameters for regular values of spectral parameters. The existence and uniqueness of redefined functions have been justified by solving the systems of two countable systems of nonlinear integral equations. The results are formulated as a theorem.

Keywords: integro-differential equation; mixed type equation; small parameter; spectral parameters; Caputo operators of different fractional orders; inverse problem; one value solvability

1. Introduction

Fractional calculus plays an important role in the mathematical modeling of many natural and engineering processes (see [1]). We can gladly refer to many examples of applied research works, where fractional integro-differential operators are successfully and widely used. For example, in [2] some applications of basic problems in continuum and statistical mechanics are considered. In [3], the mathematical problems of an Ebola epidemic model by fractional order equations are studied. In [4,5], fractional models of the dynamics of tuberculosis infection and novel coronavirus (nCoV-2019) are studied, respectively. The construction of various models for studying problems of theoretical physics by the aid of fractional calculus is described in [6] (vol. 4, 5), [7,8]. A specific physical interpretation of the fractional derivatives, describing the random motion of a particle moving on the real line at Poisson-paced times with finite velocity, is given in [9]. A detailed review of the applications of fractional calculus in solving practical problems is given in [6] (vol. 6–8), [10]. More detailed information, as well as a bibliography related to the theory of fractional integro-differentiation and fractional derivatives, can also be found in [11–18].

We also note the special role of generalized special functions, such as polynomials, in solving fractional differential equations. In [19], using Hermite polynomials of higher and fractional order, some operational techniques to find general solutions of extended forms to d'Alembert and Fourier equations. In [20], the solutions of various generalized forms of the Heat Equation, by means of different tools ranging from the use of Hermite–Kampé de Fériet polynomials of higher and fractional order to operational techniques, are discussed. In [21], the combined use of integral transforms and special polynomials provides a powerful tool to deal with fractional derivatives and integrals. The real need to know the properties of such special functions in solving direct and inverse problems for fractional partial differential equations has been shown in [22].

Applications for equations of mixed type are studied in the works of many researchers. For example, in [23], an example of gas motion in a channel surrounded by a porous medium was studied, with the gas motion in a channel being described by a wave equation, while—outside the channel—a diffusion equation was posed. In [24], a problem related to the propagation of electric oscillations in compound lines, when the losses on a semi-infinite line were neglected and the rest of the line was treated as a cable with no leaks, was investigated. This reduced the problem under consideration to a mixed parabolic–hyperbolic type equation. In [25], a hyperbolic–parabolic system, in relation to pulse combustion, is investigated. Mixed type fractional differential equations are studied in many works by scientists—particularly in [26–35].

The theories of integral and integro-differential equations are important in studying the large directions of the general theory of equations of mathematical physics. The presence of an integral term in differential equations of the first and second order has an important role in the theory of dynamical systems of automatic control [36,37]. Boundary value problems for integro-differential equations with spectral parameters have singularities in studying the questions of one-value solvability [38,39]. Mixed type integer order integro-differential equations with degenerate kernels and spectral parameters are studied in [40,41].

To find the solutions of direct mixed and boundary value problems of mathematical physics, it is required to set the coefficients of the equation, the boundary of the domain under consideration, and the initial and boundary data. It usually happens that, in solving the applied problems experimentally, the quantitative characteristics of the object under study are not available for direct observation, or it is impossible to carry out the experiment itself for one reason or another. Then, in practice, the researchers can obtain some indirect information and draw a conclusion about the properties of the studied object. This information is determined by the nature of the object under study and here requires mathematical processing and the interpretation of research results. Nonlocal integral conditions often arise when the experiment gives averaged information about this object. When the structure of the mathematical model of the studying process is known, the problem of redefining the mathematical model is posed. Such problems belong to the class of inverse problems. By inverse problems we mean problems whose solution consists of determining the parameters of a model based on the available observation results and other experimental information. Inverse problems for equations of mixed type are studied relatively rarely due to the complexity of the studying process.

In the present paper, we study the questions of the one-value solvability of an inverse boundary value problem for a mixed type integro-differential equation with Caputo operators of different fractional orders and spectral parameters in a multidimensional rectangular domain.

The rest of this paper is organized as follows. In Section 2, we state the problem, which we will investigate in this work. Section 3 is devoted to formally expanding the solution of the direct problem into Fourier series. In Section 4, we formally determine the redefinition functions. Section 5 contains the proof of existence and uniqueness of Fourier coefficients of redefinition functions from a countable system of nonlinear integral equations. Section 6 is devoted to the justification of convergence and the possibility of the term by term differentiation of the obtained Fourier series. Section 7 contains the proof of the continuous dependence of the solution on the small parameter. In the last Section 8, as a conclusion, we formulate the theorem, which we have proved in this paper.

2. Statement of the Problem

We recall that the Caputo differential operator of fractional order $m - 1 < \alpha < m$ has the form

$$cD_{at}^{\alpha} f(t) = \frac{1}{\Gamma(m - \alpha)} \int_a^t (t - s)^{m - \alpha - 1} f^{(m)}(s) \, ds,$$

where $\Gamma(z)$ is Euler gamma function.

In the multidimensional domain $\Omega = \{-T < t < T, 0 < x_1, \ldots, x_m < l\}$, a mixed type integro-differential equation of the following form is considered:

$$A_\varepsilon(U) - B_\omega(U) = \begin{cases} \nu \int_0^T K_1(t, s) U(s, x) \, ds + F_1(t, x), & t > 0, \\ \nu \int_{-T}^0 K_2(t, s) U(s, x) \, ds + F_2(t, x), & t < 0, \end{cases} \quad (1)$$

where

$$F_i(t, x) = k_i(t) \left[g_i(x) + f_i \left(x, \int_{\Omega_l^m} \Theta_i(y) g_i(y) \, dy \right) \right], \quad i = 1, 2,$$

$$A_\varepsilon(U) = \frac{1 + \text{sgn}(t)}{2} \left[cD_{0t}^{\alpha_1} - \varepsilon \sum_{i=1}^m \frac{\partial^2}{\partial x_i \partial x_i} cD_{0t}^{\beta_1} \right] U(t, x)$$

$$+ \frac{1 - \text{sgn}(t)}{2} \left[cD_{0t}^{\alpha_2} - \varepsilon \sum_{i=1}^m \frac{\partial^2}{\partial x_i \partial x_i} cD_{0t}^{\beta_2} \right] U(t, x),$$

$$B_\omega(U) = \begin{cases} \sum_{i=1}^m U_{x_i x_i}, & t > 0, \\ \omega^2 \sum_{i=1}^m U_{x_i x_i}, & t < 0, \end{cases}$$

T and l are given as positive real numbers, ω is a positive spectral parameter, ε is a positive small parameter, ν is a real non-zero spectral parameter, $0 \neq K_j(t, s) = a_j(t) b_j(s)$, $a_j(t) \in C^2[-T; T]$, $b_j(s) \in C[-T; T]$, $f_i \in C_x^2(\Omega_l^m \times \mathbb{R})$, $\int_{\Omega_l^m} |\Theta_i(y)| \, dy < \infty$, $\int_{\Omega_l^m} |\Theta_i(y)| \, dy = \int_0^l \ldots \int_0^l |\Theta_i(y)| \, dy_1 \cdot \ldots \cdot dy_m$, $i, j = 1, 2$, $k_1(t) \in C^2[0; T]$, $k_2(t) \in C^2[-T; 0]$, while $g_1(x)$ and $g_2(x)$ are redefinition functions, $\mathbb{R} \equiv (-\infty; \infty)$, $x \in \Omega_l^m \equiv [0; l]^m$, $0 < \beta_1 < \alpha_1 \leq 1$, $1 < \beta_2 < \alpha_2 \leq 2$.

Problem. Find in the domain Ω a triple of unknown function

$$U(t, x) \in C(\overline{\Omega}) \cap C^{0,1}(\Omega') \cap C^{\alpha_1, 2}(\Omega_+) \cap C^{\alpha_2, 2}(\Omega_-) \cap C_{t,x}^{\alpha_1+2}(\Omega_+) \cap C_{t,x}^{\alpha_2+2}(\Omega_-)$$

$$\cap C_{t, x_1, x_2, \ldots, x_m}^{\alpha_1+2+0+\ldots+0}(\Omega_+) \cap C_{t, x_1, x_2, \ldots, x_m}^{\alpha_2+2+0+\ldots+0}(\Omega_-) \cap C_{t, x_1, x_2, x_3, \ldots, x_m}^{\alpha_1+0+2+0+\ldots+0}(\Omega_+) \quad (2)$$

$$\cap C_{t, x_1, x_2, x_3, \ldots, x_m}^{\alpha_2+0+2+0+\ldots+0}(\Omega_-) \cap \ldots \cap C_{t, x_1, \ldots, x_{m-1}, x_m}^{\alpha_1+0+\ldots+0+2}(\Omega_+) \cap C_{t, x_1, \ldots, x_{m-1}, x_m}^{\alpha_2+0+\ldots+0+2}(\Omega_-)$$

and redefinition functions $g_i(x) \in C(\Omega_l^m)$, $i = 1, 2$, satisfying the mixed integro-differential Equation (1) and the following boundary conditions

$$U(-T, x) = \varphi_1(x), \quad {}_cD_{0t}^\theta U(-T, x) = \varphi_2(x), \quad x \in \Omega_l^m, \tag{3}$$

$$U(t, 0) = U(t, l) = 0, \quad -T < t < T \tag{4}$$

and additional conditions

$$\int_0^T \Phi_1(t) U(t, x) = \psi_1(x), \quad x \in \Omega_l^m, \tag{5}$$

$$\int_{-T}^0 \Phi_2(t) U(t, x) = \psi_2(x), \quad x \in \Omega_l^m, \tag{6}$$

where $0 < \theta < 1$, $\varphi_i(x)$, $\psi_i(x)$ are given smooth functions, $\varphi_i(0) = \varphi_i(l) = 0$, $\psi_i(0) = \psi_i(l) = 0$, $i = 1, 2$, $C^r(\Omega)$ is a class of functions $U(t, x_1, \ldots, x_m)$ with continuous derivatives $\frac{\partial^r U}{\partial t^r}, \frac{\partial^r U}{\partial x_1^r}, \ldots, \frac{\partial^r U}{\partial x_m^r}$ in Ω, $C_{t,x}^{r,s}(\Omega)$ is a class of functions $U(t, x_1, \ldots, x_m)$ with continuous derivatives $\frac{\partial^r U}{\partial t^r}, \frac{\partial^s U}{\partial x_1^s}, \ldots, \frac{\partial^s U}{\partial x_m^s}$ in Ω, $C_{t,x_1,x_2,\ldots,x_m}^{r+r+0+\ldots+0}(\Omega)$ is a class of functions $U(t, x_1, \ldots, x_m)$ with continuous derivative $\frac{\partial^{2r} U}{\partial t^r \partial x_1^r}$ in Ω, ..., $C_{t,x_1,\ldots,x_{m-1},x_m}^{r+0+\ldots+0+r}(\Omega)$ is a class of functions $U(t, x_1, \ldots, x_m)$ with continuous derivative $\frac{\partial^{2r} U}{\partial t^r \partial x_m^r}$ in Ω, r, s are positive real numbers, $\overline{\Omega} = \{-T \leq t \leq T, \ x \in \Omega_l^m\}$, $\Omega' = \Omega \cup \{x_1, \ldots, x_m = 0\} \cup \{x_1, \ldots, x_m = l\}$, $\Omega_- = \{-T < t < 0, \ 0 < x_1, \ldots, x_m < l\}$, $\Omega_+ = \{0 < t < T, \ 0 < x_1, \ldots, x_m < l\}$.

3. Expansion of the Solution of the Direct Problem (1)–(4) into Fourier Series

Our investigation is based on the application of sine Fourier series to the mixed type integro-differential Equation (1) of the complicated form. Hence, the solution of the mixed integro-differential Equation (1) in domain Ω is sought in the form of the following Fourier series

$$U(t, x) = \sum_{n_1, \ldots, n_m = 1}^\infty u_{n_1, \ldots, n_m}^\pm(t) \vartheta_{n_1, \ldots, n_m}(x), \tag{7}$$

where

$$u_{n_1, \ldots, n_m}^\pm(t) = \begin{cases} u_{n_1, \ldots, n_m}^+(t) = \int_{\Omega_l^m} U(t, x) \vartheta_{n_1, \ldots, n_m}(x) \, dx, & t > 0, \\ u_{n_1, \ldots, n_m}^-(t) = \int_{\Omega_l^m} U(t, x) \vartheta_{n_1, \ldots, n_m}(x) \, dx, & t < 0, \end{cases} \tag{8}$$

$$\int_{\Omega_l^m} U(t, x) \vartheta_{n_1, \ldots, n_m}(x) \, dx = \int_0^l \ldots \int_0^l U(t, x) \vartheta_{n_1, \ldots, n_m}(x) \, dx_1 \cdot \ldots \cdot dx_m,$$

$$\vartheta_{n_1, \ldots, n_m}(x) = \left(\sqrt{\frac{2}{l}}\right)^m \sin \frac{\pi n_1}{l} x_1 \cdot \ldots \cdot \sin \frac{\pi n_m}{l} x_m, \quad n_1, \ldots, n_m = 1, 2, \ldots$$

In this order, we also suppose that the redefinition functions and nonlinear functions on the right-hand side of the integro-differential Equation (1) are representable as the following Fourier series

$$g_i(x) = \sum_{n_1, \ldots, n_m = 1}^\infty g_{i n_1, \ldots, n_m} \vartheta_{n_1, \ldots, n_m}(x), \quad f_i(x, V_i) = \sum_{n_1, \ldots, n_m = 1}^\infty f_{i n_1, \ldots, n_m}(V_i) \vartheta_{n_1, \ldots, n_m}(x), \tag{9}$$

where

$$g_{i\,n_1,...,n_m} = \int_{\Omega_l^m} g_i(x)\,\vartheta_{n_1,...,n_m}(x)\,dx, \quad f_{i\,n_1,...,n_m}(V_i) = \int_{\Omega_l^m} f_i(y, V_i)\,\vartheta_{n_1,...,n_m}(y)\,d\underline{y},$$

$$f_i(y, V_i) = f_i\left(y, \int_{\Omega_l^m} \Theta_i(z)\,g_i(z)\,dz\right), \quad i = 1, 2.$$

Substituting series (7) and (9) into mixed Equation (1), we obtain two fractional countable systems of ordinary integro-differential equations

$$
{}_cD_{0t}^{\alpha_1}u_{n_1,...,n_m}^+(t) + \varepsilon\,\mu_{n_1,...,n_m}^2\,{}_cD_{0t}^{\beta_1}u_{n_1,...,n_m}^+(t) + \mu_{n_1,...,n_m}^2\,u_{n_1,...,n_m}^+(t)
$$
$$
= \nu\int_0^T a_1(t)\,b_1(s)\,u_{n_1,...,n_m}^+(s)\,ds + F_{1\,n_1,...,n_m}(t), \quad t > 0,
\tag{10}
$$

$$
{}_cD_{0t}^{\alpha_2}u_{n_1,...,n_m}^-(t) + \varepsilon\,\mu_{n_1,...,n_m}^2\,{}_cD_{0t}^{\beta_2}u_{n_1,...,n_m}^-(t) + \mu_{n_1,...,n_m}^2\,\omega^2 u_{n_1,...,n_m}^-(t)
$$
$$
= \nu\int_{-T}^0 a_2(t)\,b_2(s)\,u_{n_1,...,n_m}^-(s)\,ds + F_{2\,n_1,...,n_m}(t), \quad t < 0,
\tag{11}
$$

where $\mu_{n_1,...,n_m} = \frac{\pi}{T}\sqrt{n_1^2 + \ldots + n_m^2}$,

$$F_{i\,n_1,...,n_m}(t) = k_i(t)\,[g_{i\,n_1,...,n_m} + f_{i\,n_1,...,n_m}(V_i)], \quad i = 1, 2. \tag{12}$$

We use the method of degenerate kernels. In this order, by the aid of designations

$$\tau_{n_1,...,n_m}^+ = \int_0^T b_1(s)\,u_{n_1,...,n_m}^+(s)\,ds, \tag{13}$$

$$\tau_{n_1,...,n_m}^- = \int_{-T}^0 b_2(s)\,u_{n_1,...,n_m}^-(s)\,ds \tag{14}$$

we present the countable systems of ordinary integro-differential Equations (10) and (11) as follows

$$
{}_cD_{0t}^{\alpha_1}u_{n_1,...,n_m}^+(t) + \varepsilon\,\mu_{n_1,...,n_m}^2\,{}_cD_{0t}^{\beta_1}u_{n_1,...,n_m}^+(t) + \mu_{n_1,...,n_m}^2\,u_{n_1,...,n_m}^+(t)
$$
$$
= \nu\,a_1(t)\,\tau_{n_1,...,n_m}^+ + F_{1\,n_1,...,n_m}(t), \quad t > 0,
\tag{15}
$$

$$
{}_cD_{0t}^{\alpha_2}u_{n_1,...,n_m}^-(t) + \varepsilon\,\mu_{n_1,...,n_m}^2\,{}_cD_{0t}^{\beta_2}u_{n_1,...,n_m}^-(t) + \mu_{n_1,...,n_m}^2\,\omega^2 u_{n_1,...,n_m}^-(t)
$$
$$
= \nu\,a_2(t)\,\tau_{n_1,...,n_m}^- + F_{2\,n_1,...,n_m}(t), \quad t < 0.
\tag{16}
$$

The solutions of the countable systems of differential Equations (15) and (16), satisfying conditions

$$u_{n_1,...,n_m}^+(0) = C_{1\,n_1,...,n_m}^+, \quad u_{n_1,...,n_m}^-(0) = C_{1\,n_1,...,n_m}^-, \quad \frac{d}{dt}u_{n_1,...,n_m}^-(0) = C_{2\,n_1,...,n_m}^-$$

have the following form:

$$u_{n_1,...,n_m}^+(t) = \nu\,\tau_{n_1,...,n_m}^+\,\Psi_{11\,n_1,...,n_m}(t,\varepsilon) + \Psi_{12\,n_1,...,n_m}(t,\varepsilon) + C_{1\,n_1,...,n_m}^+\,\Psi_{13\,n_1,...,n_m}(t,\varepsilon), \quad t > 0, \tag{17}$$

$$u^-_{n_1,...,n_m}(t) = \nu \tau^-_{n_1,...,n_m} \Psi_{21\, n_1,...,n_m}(t,\varepsilon,\omega) + \Psi_{22\, n_1,...,n_m}(t,\varepsilon,\omega)$$
$$+ C^+_{1\, n_1,...,n_m} \Psi_{23\, n_1,...,n_m}(t,\varepsilon,\omega) - C^-_{2\, n_1,...,n_m} \Psi_{24\, n_1,...,n_m}(t,\varepsilon,\omega), \quad t < 0, \tag{18}$$

where $C^+_{1\, n_1,...,n_m}$, $C^-_{i\, n_1,...,n_m}$, $(i = 1, 2)$ are for unknown constants to be uniquely determined,

$$\Psi_{11\, n_1,...,n_m}(t,\varepsilon) = \int_0^t a_1(t-s) s^{\alpha_1-1} E_{(\alpha_1-\beta_1,\alpha_1),\alpha_1}\left(-\varepsilon \mu^2_{n_1,...,n_m} s^{\alpha_1-\beta_1}, -\mu^2_{n_1,...,n_m} s^{\alpha_1}\right) ds,$$

$$\Psi_{12\, n_1,...,n_m}(t,\varepsilon) = \int_0^t F_{1\, n_1,...,n_m}(t-s) s^{\alpha_1-1} E_{(\alpha_1-\beta_1,\alpha_1),\alpha_1}\left(-\varepsilon \mu^2_{n_1,...,n_m} s^{\alpha_1-\beta_1}, -\mu^2_{n_1,...,n_m} s^{\alpha_1}\right) ds,$$

$$\Psi_{13\, n_1,...,n_m}(t,\varepsilon) = E_{(\alpha_1-\beta_1,\alpha_1),1}\left(-\varepsilon \mu^2_{n_1,...,n_m} t^{\alpha_1-\beta_1}, -\mu^2_{n_1,...,n_m} t^{\alpha_1}\right),$$

$$\Psi_{21\, n_1,...,n_m}(t,\varepsilon,\omega) = \int_t^0 a_2(s-t)(-s)^{\alpha_2-1} \Psi_{25\, n_1,...,n_m}(t,\varepsilon,\omega) ds,$$

$$\Psi_{22\, n_1,...,n_m}(t,\varepsilon,\omega) = \int_t^0 F_{2\, n_1,...,n_m}(s-t)(-s)^{\alpha_2-1} \Psi_{25\, n_1,...,n_m}(t,\varepsilon,\omega) ds,$$

$$\Psi_{23\, n_1,...,n_m}(t,\varepsilon,\omega) = E_{(\alpha_2-\beta_2,\alpha_2),1}\left(-\varepsilon \mu^2_{n_1,...,n_m}(-t)^{\alpha_2-\beta_2}, -\mu^2_{n_1,...,n_m} \omega^2(-t)^{\alpha_2}\right),$$

$$\Psi_{24\, n_1,...,n_m}(t,\varepsilon,\omega) = t\, E_{(\alpha_2-\beta_2,\alpha_2),2}\left(-\varepsilon \mu^2_{n_1,...,n_m}(-t)^{\alpha_2-\beta_2}, -\mu^2_{n_1,...,n_m} \omega^2(-t)^{\alpha_2}\right),$$

$$\Psi_{25\, n_1,...,n_m}(t,\varepsilon,\omega) = E_{(\alpha_2-\beta_2,\alpha_2),\alpha_2}\left(-\varepsilon \mu^2_{n_1,...,n_m}(-s)^{\alpha_2-\beta_2}, -\mu^2_{n_1,...,n_m} \omega^2(-s)^{\alpha_2}\right),$$

The function $E_{(\alpha,\beta),\gamma}(z_1, z_2)$ is a Mittag–Leffler function of two variables:

$$E_{(\alpha,\beta),\gamma}(z_1, z_2) = \sum_{m_1, m_2 = 0}^{\infty} \frac{z_1^{m_1} z_2^{m_2}}{\Gamma(\gamma + \alpha m_1 + \beta m_2)},$$

where z_i, α, β, $\gamma \in \mathbb{C}$, $\text{Re}(\alpha) > 0$, $\text{Re}(\beta) > 0$.

From the statement of the problem (properties in (2)), it follows that the continuous conjugation condition is fulfilled for the main unknown function: $U(0+0, x) = U(0-0, x)$. Therefore, by taking Formula (6) into account, we have the conditions for Fourier coefficients of the main unknown function

$$\begin{aligned} u^+_{n_1,...,n_m}(0+0) &= \int_{\Omega^m_l} U(0+0, x)\, \vartheta_{n_1,...,n_m}(x)\, dx \\ &= \int_{\Omega^m_l} U(0-0, x)\, \vartheta_{n_1,...,n_m}(x)\, dx = u^-_{n_1,...,n_m}(0-0). \end{aligned} \tag{19}$$

We put

$$\varphi_{i\, n_1,...,n_m} = \int_{\Omega^m_l} \varphi_i(x)\, \vartheta_{n_1,...,n_m}(x)\, dx, \quad i = 1, 2.$$

Then, taking (8) into account, from the conditions in (3), we obtain

$$u^-_{n_1,...,n_m}(-T) = \int_{\Omega^m_l} U(-T, x)\, \vartheta_{n_1,...,n_m}(x)\, dx = \int_{\Omega^m_l} \varphi_1(x)\, \vartheta_{n_1,...,n_m}(x)\, dx = \varphi_{1\, n_1,...,n_m}, \tag{20}$$

$$cD_{0t}^\theta u_{n_1,\ldots,n_m}^-(-T) = \int_{\Omega_l^m} cD_{0t}^\theta U(-T,x)\,\vartheta_{n_1,\ldots,n_m}(x)\,dx$$
$$= \int_{\Omega_l^m} \varphi_2(x)\,\vartheta_{n_1,\ldots,n_m}(x)\,dx = \varphi_{2\,n_1,\ldots,n_m}. \tag{21}$$

By the aid of the continuous conjugation condition (19) from (17) and (18), we have the relation that $C_{1\,n_1,\ldots,n_m}^+ = C_{1\,n_1,\ldots,n_m}^-$. To find the unknown coefficients of the integration $C_{1\,n_1,\ldots,n_m}^-$ and $C_{2\,n_1,\ldots,n_m}^-$ in (18), we use the conditions (20) and (21) and deduce the following system of linear algebraic equations:

$$\begin{cases} \nu\tau_{n_1,\ldots,n_m}^- \Psi_{21\,n_1,\ldots,n_m}(-T,\varepsilon,\omega) + \Psi_{22\,n_1,\ldots,n_m}(-T,\varepsilon,\omega) + \\ +C_{1\,n_1,\ldots,n_m}^- \Psi_{23\,n_1,\ldots,n_m}(-T,\varepsilon,\omega) - C_{2\,n_1,\ldots,n_m}^- \Psi_{24\,n_1,\ldots,n_m}(-T,\varepsilon,\omega) = \varphi_{1\,n_1,\ldots,n_m}, \\ \nu\tau_{n_1,\ldots,n_m}^- D_{0t}^\theta \Psi_{21\,n_1,\ldots,n_m}(-T,\varepsilon,\omega) + D_{0t}^\theta \Psi_{22\,n_1,\ldots,n_m}(-T,\varepsilon,\omega) + \\ +C_{1\,n_1,\ldots,n_m}^- D_{0t}^\theta \Psi_{23\,n_1,\ldots,n_m}(-T,\varepsilon,\omega) - C_{2\,n_1,\ldots,n_m}^- D_{0t}^\theta \Psi_{24\,n_1,\ldots,n_m}(-T,\varepsilon,\omega) = \varphi_{2\,n_1,\ldots,n_m}, \end{cases} \tag{22}$$

where by $D_{0t}^\theta \Psi(-T)$ is denoted $D_{0t}^\theta \Psi(t)|_{t=-T}$. We assume that

$$\sigma_{n_1,\ldots,n_m}(\omega) = \Psi_{24\,n_1,\ldots,n_m}(-T,\varepsilon,\omega) \cdot D_{0t}^\theta \Psi_{23\,n_1,\ldots,n_m}(-T,\varepsilon,\omega)$$
$$-\Psi_{23\,n_1,\ldots,n_m}(-T,\varepsilon,\omega) \cdot D_{0t}^\theta \Psi_{24\,n_1,\ldots,n_m}(-T,\varepsilon,\omega) \neq 0. \tag{23}$$

If the condition (23) is fulfilled, then the system (22) with respect to $C_{1\,n_1,\ldots,n_m}^-$ and $C_{2\,n_1,\ldots,n_m}^-$ is uniquely solvable. By solving this system (22), we arrive at the following presentations for these unknown coefficients

$$C_{1\,n_1,\ldots,n_m}^- = \frac{1}{\sigma_{n_1,\ldots,n_m}(\omega)}$$
$$\times \left[\varphi_{1\,n_1,\ldots,n_m} D_{0t}^\theta \Psi_{24\,n_1,\ldots,n_m}(-T,\varepsilon,\omega) + \varphi_{2\,n_1,\ldots,n_m} \Psi_{24\,n_1,\ldots,n_m}(-T,\varepsilon,\omega) - \nu\tau_{n_1,\ldots,n_m}^-\right.$$
$$\times \left(\Psi_{24\,n_1,\ldots,n_m}(-T,\varepsilon,\omega) D_{0t}^\theta \Psi_{21\,n_1,\ldots,n_m}(-T,\varepsilon,\omega) - \Psi_{21\,n_1,\ldots,n_m}(-T,\varepsilon,\omega) D_{0t}^\theta \Psi_{24\,n_1,\ldots,n_m}(-T,\varepsilon,\omega)\right)$$
$$\left. + \Psi_{22\,n_1,\ldots,n_m}(-T,\varepsilon,\omega) D_{0t}^\theta \Psi_{24\,n_1,\ldots,n_m}(-T,\varepsilon,\omega) - \Psi_{24\,n_1,\ldots,n_m}(-T,\varepsilon,\omega) D_{0t}^\theta \Psi_{22\,n_1,\ldots,n_m}(-T,\varepsilon,\omega)\right],$$

$$C_{2\,n_1,\ldots,n_m}^- = \frac{1}{\sigma_{n_1,\ldots,n_m}(\omega)}$$
$$\times \left[\varphi_{1\,n_1,\ldots,n_m} D_{0t}^\theta \Psi_{23\,n_1,\ldots,n_m}(-T,\varepsilon,\omega) + \varphi_{2\,n_1,\ldots,n_m} \Psi_{23\,n_1,\ldots,n_m}(-T,\varepsilon,\omega) - \nu\tau_{n_1,\ldots,n_m}^-\right.$$
$$\times \left(\Psi_{23\,n_1,\ldots,n_m}(-T,\varepsilon,\omega) D_{0t}^\theta \Psi_{21\,n_1,\ldots,n_m}(-T,\varepsilon,\omega) - \Psi_{21\,n_1,\ldots,n_m}(-T,\varepsilon,\omega) D_{0t}^\theta \Psi_{23\,n_1,\ldots,n_m}(-T,\varepsilon,\omega)\right)$$
$$\left. + \Psi_{22\,n_1,\ldots,n_m}(-T,\varepsilon,\omega) D_{0t}^\theta \Psi_{23\,n_1,\ldots,n_m}(-T,\varepsilon,\omega) - \Psi_{23\,n_1,\ldots,n_m}(-T,\varepsilon,\omega) D_{0t}^\theta \Psi_{22\,n_1,\ldots,n_m}(-T,\varepsilon,\omega)\right]$$

By substituting these results into (18) and taking into account $C_{1\,n_1,\ldots,n_m}^+ = C_{1\,n_1,\ldots,n_m}^-$ in (17) and designation (12), we obtain the following representations for the Fourier coefficients of the main unknown functions in the positive and negative parts of the domain:

$$u_{n_1,\ldots,n_m}^+(t,\varepsilon,\omega,\nu) = [\varphi_{1\,n_1,\ldots,n_m} + \varphi_{2\,n_1,\ldots,n_m}] N_{11\,n_1,\ldots,n_m}(t,\varepsilon,\omega)$$
$$+\nu\tau_{n_1,\ldots,n_m}^+ N_{12\,n_1,\ldots,n_m}(t,\varepsilon) - \nu\tau_{n_1,\ldots,n_m}^- N_{13\,n_1,\ldots,n_m}(t,\varepsilon,\omega) + [g_{1\,n_1,\ldots,n_m} + f_{1\,n_1,\ldots,n_m}(V_1)] \tag{24}$$
$$\times N_{14\,n_1,\ldots,n_m}(t,\varepsilon) + [g_{2\,n_1,\ldots,n_m} + f_{2\,n_1,\ldots,n_m}(V_2)] N_{15\,n_1,\ldots,n_m}(t,\varepsilon,\omega),\ t>0,$$

$$u_{n_1,\ldots,n_m}^-(t,\varepsilon,\omega,\nu) = \varphi_{1\,n_1,\ldots,n_m} N_{21\,n_1,\ldots,n_m}(t,\varepsilon,\omega) + \varphi_{2\,n_1,\ldots,n_m} N_{22\,n_1,\ldots,n_m}(t,\varepsilon,\omega)$$
$$+\nu\tau_{n_1,\ldots,n_m}^- N_{23\,n_1,\ldots,n_m}(t,\varepsilon,\omega) + [g_{2\,n_1,\ldots,n_m} + f_{2\,n_1,\ldots,n_m}(V_2)] N_{24\,n_1,\ldots,n_m}(t,\varepsilon,\omega),\ t<0, \tag{25}$$

where
$$N_{11\,n_1,\ldots,n_m}(t,\varepsilon,\omega) = \frac{1}{\sigma_{n_1,\ldots,n_m}(\omega)}\Psi_{13\,n_1,\ldots,n_m}(t,\varepsilon)\Psi_{24\,n_1,\ldots,n_m}(-T,\varepsilon,\omega),$$

$$N_{12\,n_1,\ldots,n_m}(t,\varepsilon) = \Psi_{11\,n_1,\ldots,n_m}(t,\varepsilon),$$

$$N_{13\,n_1,\ldots,n_m}(t,\varepsilon,\omega) = \frac{1}{\sigma_{n_1,\ldots,n_m}(\omega)}\left[\Psi_{24\,n_1,\ldots,n_m}(-T,\varepsilon,\omega)D_{0t}^{\theta}\Psi_{21\,n_1,\ldots,n_m}(-T,\varepsilon,\omega)\right.$$

$$\left.-\Psi_{21\,n_1,\ldots,n_m}(-T,\varepsilon,\omega)D_{0t}^{\theta}\Psi_{24\,n_1,\ldots,n_m}(-T,\varepsilon,\omega)\right]\Psi_{13\,n_1,\ldots,n_m}(t,\varepsilon),$$

$$N_{14\,n_1,\ldots,n_m}(t,\varepsilon) = \overline{\Psi}_{12\,n_1,\ldots,n_m}(t,\varepsilon),$$

$$N_{15\,n_1,\ldots,n_m}(t,\varepsilon,\omega) = \frac{1}{\sigma_{n_1,\ldots,n_m}(\omega)}\left[\overline{\Psi}_{22\,n_1,\ldots,n_m}(-T,\varepsilon,\omega)D_{0t}^{\theta}\Psi_{24\,n_1,\ldots,n_m}(-T,\varepsilon,\omega)\right.$$

$$\left.-\Psi_{24\,n_1,\ldots,n_m}(-T,\varepsilon,\omega)D_{0t}^{\theta}\overline{\Psi}_{22\,n_1,\ldots,n_m}(-T,\varepsilon,\omega)\right]\Psi_{13\,n_1,\ldots,n_m}(t,\varepsilon),$$

$$N_{21\,n_1,\ldots,n_m}(t,\varepsilon,\omega) = \frac{1}{\sigma_{n_1,\ldots,n_m}(\omega)}\left[\Psi_{23\,n_1,\ldots,n_m}(t,\varepsilon,\omega)D_{0t}^{\theta}\Psi_{24\,n_1,\ldots,n_m}(-T,\varepsilon,\omega)\right.$$

$$\left.-\Psi_{24\,n_1,\ldots,n_m}(t,\varepsilon,\omega)D_{0t}^{\theta}\Psi_{23\,n_1,\ldots,n_m}(-T,\varepsilon,\omega)\right],$$

$$N_{22\,n_1,\ldots,n_m}(t,\varepsilon,\omega) = \frac{1}{\sigma_{n_1,\ldots,n_m}(\omega)}\left[\Psi_{23\,n_1,\ldots,n_m}(t,\varepsilon,\omega)\Psi_{24\,n_1,\ldots,n_m}(-T,\varepsilon,\omega)\right.$$

$$\left.+\Psi_{24\,n_1,\ldots,n_m}(t,\varepsilon,\omega)\Psi_{23\,n_1,\ldots,n_m}(-T,\varepsilon,\omega)\right],\quad N_{23\,n_1,\ldots,n_m}(t,\varepsilon,\omega) = \Psi_{21\,n_1,\ldots,n_m}(t,\varepsilon,\omega)$$

$$-\frac{1}{\sigma_{n_1,\ldots,n_m}(\omega)}\left[\Psi_{24\,n_1,\ldots,n_m}(-T,\varepsilon,\omega)D_{0t}^{\theta}\Psi_{21\,n_1,\ldots,n_m}(-T,\varepsilon,\omega)\right.$$

$$\left.-\Psi_{21\,n_1,\ldots,n_m}(-T,\varepsilon,\omega)D_{0t}^{\theta}\Psi_{24\,n_1,\ldots,n_m}(-T,\varepsilon,\omega)\right]\Psi_{23\,n_1,\ldots,n_m}(t,\varepsilon,\omega)$$

$$+\frac{1}{\sigma_{n_1,\ldots,n_m}(\omega)}\left[\Psi_{23\,n_1,\ldots,n_m}(-T,\varepsilon,\omega)D_{0t}^{\theta}\Psi_{21\,n_1,\ldots,n_m}(-T,\varepsilon,\omega)\right.$$

$$\left.-\Psi_{21\,n_1,\ldots,n_m}(-T,\varepsilon,\omega)D_{0t}^{\theta}\Psi_{23\,n_1,\ldots,n_m}(-T,\varepsilon,\omega)\right]\Psi_{24\,n_1,\ldots,n_m}(t,\varepsilon,\omega),$$

$$N_{24\,n_1,\ldots,n_m}(t,\varepsilon,\omega) = \overline{\Psi}_{22\,n_1,\ldots,n_m}(t,\varepsilon,\omega)$$

$$+\frac{1}{\sigma_{n_1,\ldots,n_m}(\omega)}\left[\overline{\Psi}_{22\,n_1,\ldots,n_m}(-T,\varepsilon,\omega)D_{0t}^{\theta}\Psi_{24\,n_1,\ldots,n_m}(-T,\varepsilon,\omega)\right.$$

$$\left.-\Psi_{24\,n_1,\ldots,n_m}(-T,\varepsilon,\omega)D_{0t}^{\theta}\overline{\Psi}_{22\,n_1,\ldots,n_m}(-T,\varepsilon,\omega)\right]\Psi_{23\,n_1,\ldots,n_m}(t,\varepsilon,\omega)$$

$$+\frac{1}{\sigma_{n_1,\ldots,n_m}(\omega)}\left[\overline{\Psi}_{22\,n_1,\ldots,n_m}(-T,\varepsilon,\omega)D_{0t}^{\theta}\Psi_{23\,n_1,\ldots,n_m}(-T,\varepsilon,\omega)\right.$$

$$\left.-\Psi_{23\,n_1,\ldots,n_m}(-T,\varepsilon,\omega)D_{0t}^{\theta}\overline{\Psi}_{22\,n_1,\ldots,n_m}(-T,\varepsilon,\omega)\right]\Psi_{24\,n_1,\ldots,n_m}(t,\varepsilon,\omega),$$

$$\overline{\Psi}_{12\,n_1,\ldots,n_m}(t,\varepsilon) = \int_0^t k_1(t-s)\,s^{\alpha_1-1}\,E_{(\alpha_1-\beta_1,\alpha_1),\alpha_1}\left(-\varepsilon\mu_{n_1,\ldots,n_m}^2 s^{\alpha_1-\beta_1},-\mu_{n_1,\ldots,n_m}^2 s^{\alpha_1}\right)ds,$$

$$\overline{\Psi}_{22\,n_1,\ldots,n_m}(t,\varepsilon,\omega) = \int_t^0 k_2(s-t)\,(-s)^{\alpha_2-1}\Psi_{25\,n_1,\ldots,n_m}(t,\varepsilon,\omega)\,ds.$$

According to the degenerate kernels method, we substitute these presentations, (24) and (25), into designations (13) and (14):

$$\tau^+_{n_1,\ldots,n_m}\left[1 - \nu \chi_{12\,n_1,\ldots,n_m}(\varepsilon,\omega)\right] + \nu \tau^-_{n_1,\ldots,n_m} \chi_{13\,n_1,\ldots,n_m}(\varepsilon,\omega)$$
$$= \left[\varphi_{1\,n_1,\ldots,n_m} + \varphi_{2\,n_1,\ldots,n_m}\right] \chi_{11\,n_1,\ldots,n_m}(\varepsilon,\omega) + \left[g_{1\,n_1,\ldots,n_m} + f_{1\,n_1,\ldots,n_m}(V_1)\right] \chi_{14\,n_1,\ldots,n_m}(\varepsilon,\omega) \quad (26)$$
$$+ \left[g_{2\,n_1,\ldots,n_m} + f_{2\,n_1,\ldots,n_m}(V_2)\right] \chi_{15\,n_1,\ldots,n_m}(\varepsilon,\omega),$$

$$\tau^-_{n_1,\ldots,n_m}\left[1 - \nu \chi_{23\,n_1,\ldots,n_m}(\varepsilon,\omega)\right] = \varphi_{1\,n_1,\ldots,n_m}\chi_{21\,n_1,\ldots,n_m}(\varepsilon,\omega) + \varphi_{2\,n_1,\ldots,n_m}\chi_{22\,n_1,\ldots,n_m}(\varepsilon,\omega)$$
$$+ \left[g_{2\,n_1,\ldots,n_m} + f_{2\,n_1,\ldots,n_m}(V_2)\right] \chi_{24\,n_1,\ldots,n_m}(\varepsilon,\omega), \quad (27)$$

where

$$\chi_{1i\,n_1,\ldots,n_m}(\varepsilon,\omega) = \int_0^T b_1(s) N_{1i\,n_1,\ldots,n_m}(s,\varepsilon,\omega)\,ds, \quad i = \overline{1,5},$$

$$\chi_{2i\,n_1,\ldots,n_m}(\varepsilon,\omega) = \int_{-T}^0 b_2(s) N_{2i\,n_1,\ldots,n_m}(s,\varepsilon,\omega)\,ds, \quad i = \overline{1,4}.$$

We solve the linear algebraic Equations (26) and (27) as a system of algebraic equations with respect to quantities $\tau^+_{n_1,\ldots,n_m}$ and $\tau^-_{n_1,\ldots,n_m}$. If the following conditions are fulfilled

$$\nu \chi_{12\,n_1,\ldots,n_m}(\varepsilon,\omega) \neq 1, \quad \nu \chi_{23\,n_1,\ldots,n_m}(\varepsilon,\omega) \neq 1, \quad (28)$$

then, from (26) and (27), we derive

$$\tau^+_{n_1,\ldots,n_m} = \varphi_{1\,n_1,\ldots,n_m} M_{11\,n_1,\ldots,n_m}(\varepsilon,\omega) + \varphi_{2\,n_1,\ldots,n_m} M_{12\,n_1,\ldots,n_m}(\varepsilon,\omega)$$
$$+ \left[g_{1\,n_1,\ldots,n_m} + f_{1\,n_1,\ldots,n_m}(V_1)\right] M_{13\,n_1,\ldots,n_m}(\varepsilon,\omega) + \left[g_{2\,n_1,\ldots,n_m} + f_{2\,n_1,\ldots,n_m}(V_2)\right] M_{14\,n_1,\ldots,n_m}(\varepsilon,\omega), \quad (29)$$

$$\tau^-_{n_1,\ldots,n_m} = \varphi_{1\,n_1,\ldots,n_m} M_{21\,n_1,\ldots,n_m}(\varepsilon,\omega) + \varphi_{2\,n_1,\ldots,n_m} M_{22\,n_1,\ldots,n_m}(\varepsilon,\omega)$$
$$+ \left[g_{2\,n_1,\ldots,n_m} + f_{2\,n_1,\ldots,n_m}(V_2)\right] M_{23\,n_1,\ldots,n_m}(\varepsilon,\omega), \quad (30)$$

where

$$M_{1i\,n_1,\ldots,n_m}(\varepsilon,\omega) = \frac{1}{1 - \nu \chi_{12\,n_1,\ldots,n_m}(\varepsilon,\omega)}\left[\chi_{1 1\,n_1,\ldots,n_m}(\varepsilon,\omega) - \nu \frac{\chi_{13\,n_1,\ldots,n_m}(\varepsilon,\omega) \chi_{2i\,n_1,\ldots,n_m}(\varepsilon,\omega)}{1 - \nu \chi_{23\,n_1,\ldots,n_m}(\varepsilon,\omega)}\right],$$

$$i = 1, 2, \quad M_{13\,n_1,\ldots,n_m}(\varepsilon,\omega) = \frac{\chi_{14\,n_1,\ldots,n_m}(\varepsilon,\omega)}{1 - \nu \chi_{12\,n_1,\ldots,n_m}(\varepsilon,\omega)},$$

$$M_{14\,n_1,\ldots,n_m}(\varepsilon,\omega) = \frac{1}{1 - \nu \chi_{12\,n_1,\ldots,n_m}(\varepsilon,\omega)}\left[\chi_{15\,n_1,\ldots,n_m}(\varepsilon,\omega) - \nu \frac{\chi_{13\,n_1,\ldots,n_m}(\varepsilon,\omega) \chi_{24\,n_1,\ldots,n_m}(\varepsilon,\omega)}{1 - \nu \chi_{23\,n_1,\ldots,n_m}(\varepsilon,\omega)}\right],$$

$$M_{2i\,n_1,\ldots,n_m}(\varepsilon,\omega) = \frac{\chi_{2i\,n_1,\ldots,n_m}(\varepsilon,\omega)}{1 - \nu \chi_{23\,n_1,\ldots,n_m}(\varepsilon,\omega)}, \quad i = 1, 2, \quad M_{23\,n_1,\ldots,n_m}(\varepsilon,\omega) = \frac{\chi_{24\,n_1,\ldots,n_m}(\varepsilon,\omega)}{1 - \nu \chi_{23\,n_1,\ldots,n_m}(\varepsilon,\omega)}.$$

Substituting presentations (29) and (30) of $\tau^{\pm}_{n_1,\ldots,n_m}$ into (24) and (25), we derive

$$u^+_{n_1,\ldots,n_m}(t,\varepsilon,\omega,\nu) = \varphi_{1\,n_1,\ldots,n_m} Q_{11\,n_1,\ldots,n_m}(t,\varepsilon,\omega,\nu) + \varphi_{2\,n_1,\ldots,n_m} Q_{12\,n_1,\ldots,n_m}(t,\varepsilon,\omega,\nu)$$
$$+ \left[g_{1\,n_1,\ldots,n_m} + f_{1\,n_1,\ldots,n_m}(V_1)\right] Q_{13\,n_1,\ldots,n_m}(t,\varepsilon,\omega,\nu) \quad (31)$$
$$+ \left[g_{2\,n_1,\ldots,n_m} + f_{2\,n_1,\ldots,n_m}(V_2)\right] Q_{14\,n_1,\ldots,n_m}(t,\varepsilon,\omega,\nu), \quad t > 0,$$

$$u^-_{n_1,\ldots,n_m}(t,\varepsilon,\omega,\nu) = \varphi_{1\,n_1,\ldots,n_m} Q_{21\,n_1,\ldots,n_m}(t,\varepsilon,\omega,\nu) + \varphi_{2\,n_1,\ldots,n_m} Q_{22\,n_1,\ldots,n_m}(t,\varepsilon,\omega,\nu)$$
$$+ \left[g_{2\,n_1,\ldots,n_m} + f_{2\,n_1,\ldots,n_m}(V_2)\right] Q_{23\,n_1,\ldots,n_m}(t,\varepsilon,\omega,\nu), \quad t < 0, \tag{32}$$

where

$$Q_{1i\,n_1,\ldots,n_m}(t,\varepsilon,\omega,\nu) = N_{11\,n_1,\ldots,n_m}(t,\varepsilon,\omega) + \nu\, N_{12\,n_1,\ldots,n_m}(t,\varepsilon,\omega)\, M_{1i\,n_1,\ldots,n_m}(\varepsilon,\omega)$$
$$- \nu\, N_{13\,n_1,\ldots,n_m}(t,\varepsilon,\omega)\, M_{2i\,n_1,\ldots,n_m}(\varepsilon,\omega), \quad i = 1,2,$$

$$Q_{13\,n_1,\ldots,n_m}(t,\varepsilon,\omega,\nu) = N_{14\,n_1,\ldots,n_m}(t,\varepsilon,\omega) + \nu\, N_{12\,n_1,\ldots,n_m}(t,\varepsilon,\omega)\, M_{13\,n_1,\ldots,n_m}(\varepsilon,\omega),$$

$$Q_{14\,n_1,\ldots,n_m}(t,\varepsilon,\omega,\nu) = N_{15\,n_1,\ldots,n_m}(t,\varepsilon,\omega) + \nu\, N_{12\,n_1,\ldots,n_m}(t,\varepsilon,\omega)\, M_{14\,n_1,\ldots,n_m}(\varepsilon,\omega)$$
$$- \nu\, N_{13\,n_1,\ldots,n_m}(t,\varepsilon,\omega)\, M_{23\,n_1,\ldots,n_m}(\varepsilon,\omega),$$

$$Q_{2i\,n_1,\ldots,n_m}(t,\varepsilon,\omega,\nu) = N_{2i\,n_1,\ldots,n_m}(t,\varepsilon,\omega) + \nu\, N_{23\,n_1,\ldots,n_m}(t,\varepsilon,\omega)\, M_{2i\,n_1,\ldots,n_m}(\varepsilon,\omega), \quad i = 1,2,$$

$$Q_{23\,n_1,\ldots,n_m}(t,\varepsilon,\omega,\nu) = N_{24\,n_1,\ldots,n_m}(t,\varepsilon,\omega) + \nu\, N_{23\,n_1,\ldots,n_m}(t,\varepsilon,\omega)\, M_{23\,n_1,\ldots,n_m}(\varepsilon,\omega).$$

Now, we substitute presentations (31) and (32) into the Fourier series (7) and obtain the following formal solution of the direct problem (1)–(4)

$$U(t,x,\varepsilon,\omega,\nu) = \sum_{n_1,\ldots,n_m=1}^{\infty} \vartheta_{n_1,\ldots,n_m}(x) \left[\varphi_{1\,n_1,\ldots,n_m} Q_{11\,n_1,\ldots,n_m}(t,\varepsilon,\omega,\nu)\right.$$
$$+ \varphi_{2\,n_1,\ldots,n_m} Q_{12\,n_1,\ldots,n_m}(t,\varepsilon,\omega,\nu) + (g_{1\,n_1,\ldots,n_m} + f_{1\,n_1,\ldots,n_m}(V_1))\, Q_{13\,n_1,\ldots,n_m}(t,\varepsilon,\omega,\nu) \tag{33}$$
$$\left. + (g_{2\,n_1,\ldots,n_m} + f_{2\,n_1,\ldots,n_m}(V_2))\, Q_{14\,n_1,\ldots,n_m}(t,\varepsilon,\omega,\nu)\right], \quad t > 0,$$

$$U(t,x,\varepsilon,\omega,\nu) = \sum_{n_1,\ldots,n_m=1}^{\infty} \vartheta_{n_1,\ldots,n_m}(x) \left[\varphi_{1\,n_1,\ldots,n_m} Q_{21\,n_1,\ldots,n_m}(t,\varepsilon,\omega,\nu) + \varphi_{2\,n_1,\ldots,n_m}\right.$$
$$\left. \times Q_{22\,n_1,\ldots,n_m}(t,\varepsilon,\omega,\nu) + (g_{2\,n_1,\ldots,n_m} + f_{2\,n_1,\ldots,n_m}(V_2))\, Q_{23\,n_1,\ldots,n_m}(t,\varepsilon,\omega,\nu)\right], \quad t < 0. \tag{34}$$

We suppose that the conditions of (23) were violated for some values of spectral parameter ω. So, we have to consider the algebraic equation with respect to spectral parameter ω

$$\sigma_{n_1,\ldots,n_m}(\omega) = \Psi_{24n_1,\ldots,n_m}(-T,\varepsilon,\omega) \cdot D^\theta_{0t} \Psi_{23n_1,\ldots,n_m}(-T,\varepsilon,\omega)$$
$$- \Psi_{23n_1,\ldots,n_m}(-T,\varepsilon,\omega) \cdot D^\theta_{0t} \Psi_{24n_1,\ldots,n_m}(-T,\varepsilon,\omega) = 0. \tag{35}$$

The set of positive solutions of this algebraic Equation (35) with respect to the spectral parameter ω, we denote by $\Im_1$. We call these values $\omega \in \Im_1$ as irregular values and, for these values, the condition (23) is violated. Another set $\Lambda_1 = (0;\infty) \setminus \Im_1$ is called the set of regular values of the spectral parameter ω and, for these regular values, the condition (23) is fulfilled.

Now, we assume that the conditions in (28) are violated $\nu\, \chi_{12\,n_1,\ldots,n_m}(\varepsilon,\omega) = 1$, $\nu\, \chi_{23\,n_1,\ldots,n_m}(\varepsilon,\omega) = 1$. Hence, we have

$$\nu_1 = \frac{1}{\chi_{12\,n_1,\ldots,n_m}(\varepsilon,\omega)}, \quad \nu_2 = \frac{1}{\chi_{23\,n_1,\ldots,n_m}(\varepsilon,\omega)}.$$

For regular values $\omega \in \Lambda_1$ there hold $\chi_{12\,n_1,\ldots,n_m}(\varepsilon,\omega) \neq 0$, $\chi_{23\,n_1,\ldots,n_m}(\varepsilon,\omega) \neq 0$. So, we denote the set $\{\nu_1,\nu_2\}$ by $\Im_2$. Then a set $\Lambda_2 = (-\infty;0) \cup (0;\infty) \setminus \Im_2$ is called the set of regular values of the spectral parameter ν. Therefore, for all values of $\nu \in \Lambda_2$, condition (28) is satisfied. We use the following notation $\aleph = \{n_1,\ldots,n_m \in \mathbb{N}; \omega \in \Lambda_1; \nu \in \Lambda_2\}$, where $\mathbb{N}$ is the set of natural numbers. This is the set on which all values of the spectral parameters ω and ν are regular. Therefore, in this case, we study the solution of the direct problem (1)–(4) in the domain Ω as Fourier series (33) and (34).

4. Redefinition Functions

Suppose that the functions $\psi_i(x)$ expand in the Fourier series

$$\psi_i(x) = \sum_{n_1,\ldots,n_m=1}^{\infty} \psi_{i\,n_1,\ldots,n_m}\,\vartheta_{n_1,\ldots,n_m}(x), \qquad (36)$$

where

$$\psi_{i\,n_1,\ldots,n_m} = \int_{\Omega_l^m} \psi_i(x)\,\vartheta_{n_1,\ldots,n_m}(x)\,dx, \quad i=1,2,\ \ n_1,\ldots,n_m=1,2,\ldots$$

By virtue of series (33), (34) and (36), we apply conditions (5) and (6):

$$\sum_{n_1,\ldots,n_m=1}^{\infty} \psi_{1\,n_1,\ldots,n_m}\,\vartheta_{n_1,\ldots,n_m}(x) = \sum_{n_1,\ldots,n_m=1}^{\infty} \vartheta_{n_1,\ldots,n_m}(x)$$

$$\times \int_0^T \Phi_1(t)\,\big[\varphi_{1\,n_1,\ldots,n_m} Q_{11\,n_1,\ldots,n_m}(t,\varepsilon,\omega,\nu) + \varphi_{2\,n_1,\ldots,n_m} Q_{12\,n_1,\ldots,n_m}(t,\varepsilon,\omega,\nu)$$

$$+ (g_{1\,n_1,\ldots,n_m} + f_{1\,n_1,\ldots,n_m}(V_1))\,Q_{13\,n_1,\ldots,n_m}(t,\varepsilon,\omega,\nu) + (g_{2\,n_1,\ldots,n_m} + f_{2\,n_1,\ldots,n_m}(V_2))\,Q_{14\,n_1,\ldots,n_m}(t,\varepsilon,\omega,\nu)\big]\,dt,$$

$$\sum_{n_1,\ldots,n_m=1}^{\infty} \psi_{2\,n_1,\ldots,n_m}\,\vartheta_{n_1,\ldots,n_m}(x) = \sum_{n_1,\ldots,n_m=1}^{\infty} \vartheta_{n_1,\ldots,n_m}(x)$$

$$\times \int_{-T}^{0} \Phi_2(t)\,\big[\varphi_{1\,n_1,\ldots,n_m} Q_{21\,n_1,\ldots,n_m}(t,\varepsilon,\omega,\nu) + \varphi_{2\,n_1,\ldots,n_m} Q_{22\,n_1,\ldots,n_m}(t,\varepsilon,\omega,\nu)$$

$$+ (g_{2\,n_1,\ldots,n_m} + f_{2\,n_1,\ldots,n_m}(V_2))\,Q_{23\,n_1,\ldots,n_m}(t,\varepsilon,\omega,\nu)\big]\,dt.$$

Hence, we obtain

$$\begin{aligned}\psi_{1\,n_1,\ldots,n_m} &= \varphi_{1\,n_1,\ldots,n_m}\,Y_{11\,n_1,\ldots,n_m}(\varepsilon,\omega,\nu) + \varphi_{2\,n_1,\ldots,n_m}\,Y_{12\,n_1,\ldots,n_m}(\varepsilon,\omega,\nu)\\ &\quad + (g_{1\,n_1,\ldots,n_m} + f_{1\,n_1,\ldots,n_m}(V_1))\,Y_{13\,n_1,\ldots,n_m}(\varepsilon,\omega,\nu)\\ &\quad + (g_{2\,n_1,\ldots,n_m} + f_{2\,n_1,\ldots,n_m}(V_2))\,Y_{14\,n_1,\ldots,n_m}(\varepsilon,\omega,\nu),\end{aligned} \qquad (37)$$

$$\begin{aligned}\psi_{2\,n_1,\ldots,n_m} &= \varphi_{1\,n_1,\ldots,n_m}\,Y_{21\,n_1,\ldots,n_m}(\varepsilon,\omega,\nu) + \varphi_{2\,n_1,\ldots,n_m}\,Y_{22\,n_1,\ldots,n_m}(\varepsilon,\omega,\nu)\\ &\quad + (g_{2\,n_1,\ldots,n_m} + f_{2\,n_1,\ldots,n_m}(V_2))\,Y_{23\,n_1,\ldots,n_m}(\varepsilon,\omega,\nu),\end{aligned} \qquad (38)$$

where

$$Y_{1\,i\,n_1,\ldots,n_m}(\varepsilon,\omega,\nu) = \int_0^T \Phi_1(t)\,Q_{1\,i\,n_1,\ldots,n_m}(t,\varepsilon,\omega,\nu)\,dt, \quad i=\overline{1,4},$$

$$Y_{2\,i\,n_1,\ldots,n_m}(\varepsilon,\omega,\nu) = \int_{-T}^{0} \Phi_2(t)\,Q_{2\,i\,n_1,\ldots,n_m}(t,\varepsilon,\omega,\nu)\,dt, \quad i=\overline{1,3}.$$

The relations of (37) and (38) we consider as a system of functional algebraic equations with respect to coefficients of redefinition functions. By solving this system, we obtain the following representations

$$\begin{aligned}g_{1\,n_1,\ldots,n_m}(\varepsilon,\omega,\nu) + f_{1\,n_1,\ldots,n_m}(V_1) &= \psi_{1\,n_1,\ldots,n_m}\Delta_{11\,n_1,\ldots,n_m}(\varepsilon,\omega,\nu) + \psi_{2\,n_1,\ldots,n_m}\Delta_{12\,n_1,\ldots,n_m}(\varepsilon,\omega,\nu)\\ &\quad + \varphi_{1\,n_1,\ldots,n_m}\Delta_{13\,n_1,\ldots,n_m}(\varepsilon,\omega,\nu) + \varphi_{2\,n_1,\ldots,n_m}\Delta_{14\,n_1,\ldots,n_m}(\varepsilon,\omega,\nu),\end{aligned} \qquad (39)$$

$$g_{2\,n_1,\ldots,n_m}(\varepsilon,\omega,\nu) + f_{2\,n_1,\ldots,n_m}(V_2) = \psi_{2\,n_1,\ldots,n_m}\Delta_{21\,n_1,\ldots,n_m}(\varepsilon,\omega,\nu) \qquad (40)$$
$$+\varphi_{1\,n_1,\ldots,n_m}\Delta_{22\,n_1,\ldots,n_m}(\varepsilon,\omega,\nu) + \varphi_{2\,n_1,\ldots,n_m}\Delta_{23\,n_1,\ldots,n_m}(\varepsilon,\omega,\nu),$$

where

$$\Delta_{11\,n_1,\ldots,n_m}(\varepsilon,\omega,\nu) = \left(Y_{13\,n_1,\ldots,n_m}(\varepsilon,\omega,\nu)\right)^{-1},$$

$$\Delta_{12\,n_1,\ldots,n_m}(\varepsilon,\omega,\nu) = -Y_{23\,n_1,\ldots,n_m}(\varepsilon,\omega,\nu)\left(Y_{13\,n_1,\ldots,n_m}(\varepsilon,\omega,\nu)\right)^{-1},$$

$$\Delta_{13\,n_1,\ldots,n_m}(\varepsilon,\omega,\nu) = [-Y_{11\,n_1,\ldots,n_m}(\varepsilon,\omega,\nu) + Y_{21\,n_1,\ldots,n_m}(\varepsilon,\omega,\nu)Y_{23\,n_1,\ldots,n_m}(\varepsilon,\omega,\nu)]\left(Y_{13\,n_1,\ldots,n_m}(\varepsilon,\omega,\nu)\right)^{-1},$$

$$\Delta_{14\,n_1,\ldots,n_m}(\varepsilon,\omega,\nu) = [-Y_{12\,n_1,\ldots,n_m}(\varepsilon,\omega,\nu) + Y_{22\,n_1,\ldots,n_m}(\varepsilon,\omega,\nu)Y_{23\,n_1,\ldots,n_m}(\varepsilon,\omega,\nu)]\left(Y_{13\,n_1,\ldots,n_m}(\varepsilon,\omega,\nu)\right)^{-1},$$

$$\Delta_{21\,n_1,\ldots,n_m}(\varepsilon,\omega,\nu) = \left(Y_{23\,n_1,\ldots,n_m}(\varepsilon,\omega,\nu)\right)^{-1},$$

$$\Delta_{22\,n_1,\ldots,n_m}(\varepsilon,\omega,\nu) = -Y_{21\,n_1,\ldots,n_m}(\varepsilon,\omega,\nu)\left(Y_{23\,n_1,\ldots,n_m}(\varepsilon,\omega,\nu)\right)^{-1},$$

$$\Delta_{23\,n_1,\ldots,n_m}(\varepsilon,\omega,\nu) = -Y_{22\,n_1,\ldots,n_m}(\varepsilon,\omega,\nu)\left(Y_{23\,n_1,\ldots,n_m}(\varepsilon,\omega,\nu)\right)^{-1}.$$

We rewrite Formulas (39) and (40) in the form of countable systems of nonlinear integral equations (CSNIE)

$$g_{i\,n_1,\ldots,n_m}(\varepsilon,\omega,\nu) = I_i\left(g_{i\,n_1,\ldots,n_m}\right) \equiv c_{i\,n_1,\ldots,n_m}(\varepsilon,\omega,\nu) \qquad (41)$$
$$-\int_{\Omega_l^m} f_i\left(y,\int_{\Omega_l^m}\Theta_i(z)\sum_{n_1,\ldots,n_m=1}^{\infty}g_{i\,n_1,\ldots,n_m}(\varepsilon,\omega,\nu)\vartheta_{n_1,\ldots,n_m}(z)\,dz\right)\vartheta_{n_1,\ldots,n_m}(y)\,dy,\; i=1,2,$$

where

$$c_{1\,n_1,\ldots,n_m}(\varepsilon,\omega,\nu) = \psi_{1\,n_1,\ldots,n_m}\Delta_{11\,n_1,\ldots,n_m}(\varepsilon,\omega,\nu) + \psi_{2\,n_1,\ldots,n_m}\Delta_{12\,n_1,\ldots,n_m}(\varepsilon,\omega,\nu)$$
$$+\varphi_{1\,n_1,\ldots,n_m}\Delta_{13\,n_1,\ldots,n_m}(\varepsilon,\omega,\nu) + \varphi_{2\,n_1,\ldots,n_m}\Delta_{14\,n_1,\ldots,n_m}(\varepsilon,\omega,\nu),$$
$$c_{2\,n_1,\ldots,n_m}(\varepsilon,\omega,\nu) = \psi_{2\,n_1,\ldots,n_m}\Delta_{21\,n_1,\ldots,n_m}(\varepsilon,\omega,\nu)$$
$$+\varphi_{1\,n_1,\ldots,n_m}\Delta_{22\,n_1,\ldots,n_m}(\varepsilon,\omega,\nu) + \varphi_{2\,n_1,\ldots,n_m}\Delta_{23\,n_1,\ldots,n_m}(\varepsilon,\omega,\nu).$$

5. Unique Solvability of CSNIE (41)

We use the concepts of the following well-known Banach spaces, including a Hilbert coordinate space ℓ_2 of number sequences $\{b_{n_1,\ldots,n_m}\}_{n_1,\ldots,n_m=1}^{\infty}$ with the norm

$$\|b\|_{\ell_2} = \sqrt{\sum_{n_1,\ldots,n_m=1}^{\infty}|b_{n_1,\ldots,n_m}|^2} < \infty.$$

We also use the space $L_2(\Omega_l^m)$ of square-summable functions on the domain Ω_l^m with the norm

$$\|\vartheta(x)\|_{L_2(\Omega_l^m)} = \sqrt{\int_{\Omega_l^m}|\vartheta(x)|^2\,dx} < \infty.$$

In the process of proofing the unique solvability of CSNIE (41), we need the following conditions.

Smoothness conditions. Let functions

$$\varphi_i(x),\ \psi_i(x) \in C^2(\Omega_l^m),\ f_i\left(x, \int_{\Omega_l^m} \Theta_i(y) g_i(y) \, dy\right) \in C_x^2(\Omega_l^m \times \mathbb{R}),\ i = 1, 2$$

in the domain Ω_l^m have piecewise continuous third order derivatives.

Then, by integrating them in parts three times over all variables $x_1, x_2, \ldots, x_m$, we obtain the following formulas [41]

$$\sum_{n_1,\ldots,n_m=1}^{\infty} \left[\varphi_{i\,n_1,\ldots,n_m}^{(3m)}\right]^2 \leq \left(\frac{2}{l}\right)^m \int_{\Omega_l^m} \left[\frac{\partial^{3m} \varphi_i(x)}{\partial x_1^3 \partial x_2^3 \ldots \partial x_m^3}\right]^2 dx, \tag{42}$$

$$\sum_{n_1,\ldots,n_m=1}^{\infty} \left[\psi_{i\,n_1,\ldots,n_m}^{(3m)}\right]^2 \leq \left(\frac{2}{l}\right)^m \int_{\Omega_l^m} \left[\frac{\partial^{3m} \psi_i(x)}{\partial x_1^3 \partial x_2^3 \ldots \partial x_m^3}\right]^2 dx, \tag{43}$$

$$|\varphi_{i\,n_1,\ldots,n_m}| = \left(\frac{l}{\pi}\right)^{3m} \frac{\left|\varphi_{i\,n_1,\ldots,n_m}^{(3m)}\right|}{n_1^3 \ldots n_m^3},\ |\psi_{i\,n_1,\ldots,n_m}| = \left(\frac{l}{\pi}\right)^{3m} \frac{\left|\psi_{i\,n_1,\ldots,n_m}^{(3m)}\right|}{n_1^3 \ldots n_m^3}, \tag{44}$$

where

$$\varphi_{i\,n_1,\ldots,n_m}^{(3m)} = \int_{\Omega_l^m} \frac{\partial^{3m} \varphi_i(x)}{\partial x_1^3 \partial x_2^3 \ldots \partial x_m^3} \vartheta_{n_1,\ldots,n_m}(x) \, dx,$$

$$\psi_{i\,n_1,\ldots,n_m}^{(3m)} = \int_{\Omega_l^m} \frac{\partial^{5m} \psi_i(x)}{\partial x_1^3 \partial x_2^3 \ldots \partial x_m^3} \vartheta_{n_1,\ldots,n_m}(x) \, dx,\ i = 1, 2.$$

We obtain also that

$$|f_{i\,n_1,\ldots,n_m}(V_i)| = \left(\frac{l}{\pi}\right)^{3m} \frac{\left|f_{i\,n_1,\ldots,n_m}^{(3m)}(x, V_i)\right|}{n_1^3 \ldots n_m^3}, \tag{45}$$

$$\sum_{n_1,\ldots,n_m=1}^{\infty} \left[f_{i\,n_1,\ldots,n_m}^{(3m)}(V_i)\right]^2 \leq \left(\frac{2}{l}\right)^m \int_{\Omega_l^m} \left[\frac{\partial^{3m} f_i(x, V_i)}{\partial x_1^3 \partial x_2^3 \ldots \partial x_m^3}\right]^2 dx, \tag{46}$$

where

$$f_{i\,n_1,\ldots,n_m}^{(3m)}(V_i) = \int_{\Omega_l^m} \frac{\partial^{3m} f_i(x, V_i)}{\partial x_1^3 \partial x_2^3 \ldots \partial x_m^3} \vartheta_{n_1,\ldots,n_m}(x) \, dx,\ i = 1, 2.$$

We use also the following well known properties of the Mittag–Leffler function:

(1) For all $k > 0$, $\alpha_0, \beta_0, \gamma_0 \in (0; 2]$, $\alpha_0 \leq \beta_0 \leq \gamma_0$, $t \geq 0$ the function $t^{\beta_0-1} E_{\alpha_0,\beta_0,\gamma_0}(-kt^\alpha, -kt^\beta)$ is complete and monotonous and there holds

$$(-1)^s \left[t^{\beta_0-1} E_{(\alpha_0,\beta_0),\gamma_0}(-kt^{\alpha_0}, -kt^{\beta_0})\right]^{(s)} \geq 0,\ s = 0, 1, 2, \ldots \tag{47}$$

(2) For all $\alpha_0, \beta_0 \in (0, 2)$, $\gamma \in \mathbb{R}$ and $\arg z_1 = \pi$, there hold the following estimates

$$\left|E_{(\alpha_0,\beta_0),\gamma_0}(z_1, z_2)\right| \leq \frac{C_1}{1+|z_1|}, \tag{48}$$

$$\left|E_{(\alpha_0,\beta_0),\gamma_0}(\varepsilon_1 z_1, z_2) - E_{(\alpha_0,\beta_0),\gamma_0}(\varepsilon_2 z_1, z_2)\right| \leq |\varepsilon_1 - \varepsilon_2| \frac{C_2}{1+|z_1|}, \tag{49}$$

where $0 < C_i = $ const does not depend on z, $\varepsilon_i \in (0; \varepsilon_0)$, $0 < \varepsilon_0 = $ const, $i = 1, 2$.

According to the properties of the Mittag–Leffler function (Formulas (47) and (48)) the quantities $\Delta_{1,j\,n_1,\ldots,n_m}(\varepsilon, \omega, \nu)$ ($j = \overline{1,4}$) and $\Delta_{2,j\,n_1,\ldots,n_m}(\varepsilon, \omega, \nu)$ ($j = \overline{1,3}$) are uniformly bounded. So, for any positive integers $n_1, \ldots, n_m$, there exist finite constant numbers C_{0i} ($i = 1, 2$), by which the following estimates take place

$$\max_{n_1,\ldots,n_m \in \mathbb{N}} \max_j |\Delta_{1,j\,n_1,\ldots,n_m}(\varepsilon, \omega, \nu)| \leq C_{01}, \quad j = \overline{1,4}, \tag{50}$$

$$\max_{n_1,\ldots,n_m \in \mathbb{N}} \max_j |\Delta_{2,j\,n_1,\ldots,n_m}(\varepsilon, \omega, \nu)| \leq C_{02}, \quad j = \overline{1,3}, \tag{51}$$

where $0 < C_{0i} = $ const, $i = 1, 2$.

Lemma 1. *Suppose that the smoothness conditions are fulfilled and*

$$|f_i(x, V_{1i}) - f_i(x, V_{2i})| \leq K_{1i}(x) |V_{1i} - V_{2i}|, \quad \rho < 1,$$

where $\rho = C_{04} \gamma_3 \| \Theta_i(x) \|_{L_2(\Omega_l^m)}$, $\gamma_3 = C_{03} \left(\frac{l}{\pi}\right)^{3m} \left(\frac{2}{l}\right)^m$,

$$C_{03} = \sqrt{\sum_{n_1,\ldots,n_m=1}^{\infty} \frac{1}{n_1^6 \ldots n_m^6}} < \infty; \quad \max_i \left\| \frac{\partial^{3m} K_{1i}(x)}{\partial x_1^3 \partial x_2^3 \ldots \partial x_m^3} \right\|_{L_2(\Omega_l^m)} \leq C_{04} < \infty, \quad i = 1, 2.$$

Then, for regular values of spectral parameters ω *and* ν, *CSNIE (41) is uniquely solvable in the space* ℓ_2. *In this case, successive approximations are defined as follows:*

$$g^0_{i\,n_1,\ldots,n_m}(\varepsilon, \omega, \nu) = c_{i\,n_1,\ldots,n_m}, \quad g^{k+1}_{i\,n_1,\ldots,n_m}(\varepsilon, \omega, \nu) = I_{i\,n_1,\ldots,n_m}(g^k_i), \quad i = 1, 2. \tag{52}$$

Proof. We apply the method of successive approximations and the method of compressive mappings. We use Formulas (42)–(44) and estimates (50) and (51). By the aid of the Cauchy–Schwartz inequality and the Bessel inequality for the zeroth approximation of the coefficients of the redefinition functions from successive approximations (52), we obtain

$$\| g_1^0(\varepsilon, \omega, \nu) \|_{\ell_2} \leq \sum_{n_1,\ldots,n_m=1}^{\infty} |c_{1\,n_1,\ldots,n_m}(\varepsilon, \omega, \nu)|$$

$$\leq \sum_{n_1,\ldots,n_m=1}^{\infty} |\psi_{1\,n_1,\ldots,n_m} \Delta_{11\,n_1,\ldots,n_m}(\varepsilon, \omega, \nu)| + \sum_{n_1,\ldots,n_m=1}^{\infty} |\psi_{2\,n_1,\ldots,n_m} \Delta_{12\,n_1,\ldots,n_m}(\varepsilon, \omega, \nu)|$$

$$+ \sum_{n_1,\ldots,n_m=1}^{\infty} |\varphi_{1\,n_1,\ldots,n_m} \Delta_{13\,n_1,\ldots,n_m}(\varepsilon, \omega, \nu)| + \sum_{n_1,\ldots,n_m=1}^{\infty} |\varphi_{2\,n_1,\ldots,n_m} \Delta_{14\,n_1,\ldots,n_m}(\varepsilon, \omega, \nu)|$$

$$\leq C_{01} \left[\sum_{n_1,\ldots,n_m=1}^{\infty} |\psi_{1n_1,\ldots,n_m}| + \sum_{n_1,\ldots,n_m=1}^{\infty} |\psi_{2n_1,\ldots,n_m}| + \sum_{n_1,\ldots,n_m=1}^{\infty} |\varphi_{1n_1,\ldots,n_m}| + \sum_{n_1,\ldots,n_m=1}^{\infty} |\varphi_{2n_1,\ldots,n_m}| \right]$$

$$\leq C_{01} \left(\frac{l}{\pi}\right)^{3m} \left[\sum_{n_1,\ldots,n_m=1}^{\infty} \frac{|\psi^{(3m)}_{1n_1,\ldots,n_m}|}{n_1^3 \ldots n_m^3} + \sum_{n_1,\ldots,n_m=1}^{\infty} \frac{|\psi^{(3m)}_{2n_1,\ldots,n_m}|}{n_1^3 \ldots n_m^3} \right. \tag{53}$$

$$\left. + \sum_{n_1,\ldots,n_m=1}^{\infty} \frac{|\varphi^{(3m)}_{1n_1,\ldots,n_m}|}{n_1^3 \ldots n_m^3} + \sum_{n_1,\ldots,n_m=1}^{\infty} \frac{|\varphi^{(3m)}_{2n_1,\ldots,n_m}|}{n_1^3 \ldots n_m^3} \right]$$

$$\leq C_{01} \left(\frac{l}{\pi}\right)^{3m} \sqrt{\sum_{n_1,\ldots,n_m=1}^{\infty} \frac{1}{n_1^6 \ldots n_m^6}} \left[\| \psi_1^{(3m)} \|_{\ell_2} + \| \psi_2^{(3m)} \|_{\ell_2} + \| \varphi_1^{(3m)} \|_{\ell_2} + \| \varphi_2^{(3m)} \|_{\ell_2} \right]$$

$$\leq \gamma_1 \left[\left\| \frac{\partial^{3m} \psi_1(x)}{\partial x_1^3 \partial x_2^3 \ldots \partial x_m^3} \right\|_{L_2(\Omega_l^m)} + \left\| \frac{\partial^{3m} \psi_2(x)}{\partial x_1^3 \partial x_2^3 \ldots \partial x_m^3} \right\|_{L_2(\Omega_l^m)} \right.$$

$$\left. + \left\| \frac{\partial^{3m} \varphi_1(x)}{\partial x_1^3 \partial x_2^3 \ldots \partial x_m^3} \right\|_{L_2(\Omega_l^m)} + \left\| \frac{\partial^{3m} \varphi_2(x)}{\partial x_1^3 \partial x_2^3 \ldots \partial x_m^3} \right\|_{L_2(\Omega_l^m)} \right] < \infty,$$

where $\gamma_1 = C_{01} C_{03} \left(\frac{2}{l}\right)^m \left(\frac{1}{\pi}\right)^{3m}$, $C_{03} = \sqrt{\sum_{n_1,\ldots,n_m=1}^{\infty} \frac{1}{n_1^6 \ldots n_m^6}} < \infty$;

$$\|g_2^0(\varepsilon,\omega,v)\|_{\ell_2} \leq \sum_{n_1,\ldots,n_m=1}^{\infty} |c_{2n_1,\ldots,n_m}(\varepsilon,\omega,v)| \leq \sum_{n_1,\ldots,n_m=1}^{\infty} |\psi_{2n_1,\ldots,n_m} \Delta_{21 n_1,\ldots,n_m}(\varepsilon,\omega,v)|$$
$$+ \sum_{n_1,\ldots,n_m=1}^{\infty} |\varphi_{1 n_1,\ldots,n_m} \Delta_{22 n_1,\ldots,n_m}(\varepsilon,\omega,v)| + \sum_{n_1,\ldots,n_m=1}^{\infty} |\varphi_{2 n_1,\ldots,n_m} \Delta_{23 n_1,\ldots,n_m}(\varepsilon,\omega,v)|$$
$$\leq C_{02} \left(\frac{1}{\pi}\right)^{3m} \left[\sum_{n_1,\ldots,n_m=1}^{\infty} \frac{|\psi_{2n_1,\ldots,n_m}^{(3m)}|}{n_1^3 \ldots n_m^3} + \sum_{n_1,\ldots,n_m=1}^{\infty} \frac{|\varphi_{1n_1,\ldots,n_m}^{(3m)}|}{n_1^3 \ldots n_m^3} + \sum_{n_1,\ldots,n_m=1}^{\infty} \frac{|\varphi_{2n_1,\ldots,n_m}^{(3m)}|}{n_1^3 \ldots n_m^3}\right] \quad (54)$$
$$\leq \gamma_2 \left[\left\|\frac{\partial^{3m} \psi_2(x)}{\partial x_1^3 \partial x_2^3 \ldots \partial x_m^3}\right\|_{L_2(\Omega_l^m)} + \left\|\frac{\partial^{3m} \varphi_1(x)}{\partial x_1^3 \partial x_2^3 \ldots \partial x_m^3}\right\|_{L_2(\Omega_l^m)} + \left\|\frac{\partial^{3m} \varphi_2(x)}{\partial x_1^3 \partial x_2^3 \ldots \partial x_m^3}\right\|_{L_2(\Omega_l^m)}\right] < \infty,$$

where $\gamma_2 = C_{02} C_{03} \left(\frac{2}{l}\right)^m \left(\frac{1}{\pi}\right)^{3m}$.

By Formulas (45) and (46), using the Cauchy–Schwartz inequality and Bessel inequality for the first difference of approximation (52), we obtain

$$\|g_i^1(\varepsilon,\omega,v) - g_i^0(\varepsilon,\omega,v)\|_{\ell_2} \leq \sum_{n_1,\ldots,n_m=1}^{\infty} \left|\int_{\Omega_l^m} f_i\left(x, \int_{\Omega_l^m} \Theta_i(y) g_i^0(y,\varepsilon,\omega,v) dy\right) \vartheta_{n_1,\ldots,n_m}(x) dx\right|$$
$$\leq \left(\frac{1}{\pi}\right)^{3m} \sum_{n_1,\ldots,n_m=1}^{\infty} \frac{|f_{i n_1,\ldots,n_m}^{(3m)}(x,V_i^0)|}{n_1^3 \ldots n_m^3} \leq \gamma_3 \left\|\frac{\partial^{3m} f_i(x,V_i^0)}{\partial x_1^3 \partial x_2^3 \ldots \partial x_m^3}\right\|_{L_2(\Omega_l^m)} < \infty, \quad (55)$$

where $\gamma_3 = C_{03} \left(\frac{1}{\pi}\right)^{3m} \left(\frac{2}{l}\right)^m$, $V_i^0 = \int_{\Omega_l^m} \Theta_i(x) g_i^0(x,\varepsilon,\omega,v) dx$, $i = 1, 2$.

Analogously, by the condition of the lemma and expansion (9), using the Cauchy–Schwartz inequality and Bessel inequality for an arbitrary difference of approximation (52), we obtain

$$\left\|g_i^{k+1}(\varepsilon,\omega,v) - g_i^k(\varepsilon,\omega,v)\right\|_{\ell_2}$$
$$\leq \gamma_3 \left\|\frac{\partial^{3m}}{\partial x_1^3 \partial x_2^3 \ldots \partial x_m^3} \left|f_i(x,V_i^k) - f_i(x,V_i^{k-1})\right|\right\|_{L_2(\Omega_l^m)}$$
$$\leq \gamma_3 \int_{\Omega_l^m} |\Theta_i(y)| \cdot \left|g_i^k(y,\varepsilon,\omega,v) - g_i^{k-1}(y,\varepsilon,\omega,v)\right| dy \left\|\frac{\partial^{3m} K_{1i}(x)}{\partial x_1^3 \partial x_2^3 \ldots \partial x_m^3}\right\|_{L_2(\Omega_l^m)} \quad (56)$$
$$\leq C_{04} \gamma_3 \int_{\Omega_l^m} |\Theta_i(y)| \sum_{n_1,\ldots,n_m=1}^{\infty} \left|g_{i n_1,\ldots,n_m}^k(\varepsilon,\omega,v) - g_{i n_1,\ldots,n_m}^{k-1}(\varepsilon,\omega,v)\right| \cdot |\vartheta_{n_1,\ldots,n_m}(y)| dy$$
$$\leq \rho \left\|g_i^k(\varepsilon,\omega,v) - g_i^{k-1}(\varepsilon,\omega,v)\right\|_{\ell_2}, \quad i = 1, 2,$$

where

$$\rho = C_{04} \gamma_3 \|\Theta_i(x)\|_{L_2(\Omega_l^m)} \cdot \max_i \left\|\frac{\partial^{3m} K_{1i}(x)}{\partial x_1^3 \partial x_2^3 \ldots \partial x_m^3}\right\|_{L_2(\Omega_l^m)} \leq C_{04} < \infty, \quad i = 1, 2.$$

By the condition of the lemma, $\rho < 1$. Therefore, it follows from estimate (56) that the operators on the right-hand side of (41) are contracting. From the estimates (53)–(56), it is implied that there exists a unique pair of fixed points $\{g_{1 n_1,\ldots,n_m}(\varepsilon,\omega,v); g_{2 n_1,\ldots,n_m}(\varepsilon,\omega,v)\}$, which is a solution of CSNIE (41) in the space ℓ_2. The Lemma 1 is proved. □

6. Convergence of Fourier Series (57)

Now, we determine the redefinition functions. In this order, we substitute representations (41) into the Fourier series (9) and obtain

$$g_i(x, \varepsilon, \omega, \nu) = \sum_{n_1, \ldots, n_m = 1}^{\infty} \vartheta_{n_1, \ldots, n_m}(x) \left[c_{i\, n_1, \ldots, n_m}(\varepsilon, \omega, \nu) \right.$$
$$\left. - \int_{\Omega_l^m} f_i \left(y, \int_{\Omega_l^m} \Theta_i(z) \sum_{n_1, \ldots, n_m = 1}^{\infty} g_{i\, n_1, \ldots, n_m}(\varepsilon, \omega, \nu) \vartheta_{n_1, \ldots, n_m}(z)\, dz \right) \vartheta_{n_1, \ldots, n_m}(y)\, dy \right], \quad i = 1, 2. \tag{57}$$

We prove that the following lemma holds.

Lemma 2. *Assume that the conditions of Lemma 1 are satisfied. Then for regular values of spectral parameters ω and ν, the series (57) converge absolutely.*

Proof. We use estimates (53)–(55). Using the Cauchy–Schwartz inequality and Bessel inequality for series (57), we obtain the following estimates

$$|g_1(x, \varepsilon, \omega, \nu)| \leq \sum_{n_1, \ldots, n_m = 1}^{\infty} |\vartheta_{n_1, \ldots, n_m}(x)| \left[|c_{1\, n_1, \ldots, n_m}(\varepsilon, \omega, \nu)| \right.$$
$$\left. + \left| \int_{\Omega_l^m} f_1 \left(y, \int_{\Omega_l^m} \Theta_1(z) \sum_{n_1, \ldots, n_m = 1}^{\infty} g_{1\, n_1, \ldots, n_m}(\varepsilon, \omega, \nu) \vartheta_{n_1, \ldots, n_m}(z)\, dz \right) \vartheta_{n_1, \ldots, n_m}(y)\, dy \right| \right]$$
$$\leq C_{03} \left(\frac{2}{l} \right)^{\frac{3m}{2}} \left(\frac{l}{\pi} \right)^{3m} \left[C_{01} \left\| \psi_1^{(3m)} \right\|_{\ell_2} + C_{01} \left\| \psi_2^{(3m)} \right\|_{\ell_2} \right.$$
$$\left. + C_{01} \left\| \varphi_1^{(3m)} \right\|_{\ell_2} + C_{01} \left\| \varphi_2^{(3m)} \right\|_{\ell_2} + \left\| f_1^{(3m)}(V_1) \right\|_{\ell_2} \right] \tag{58}$$
$$\leq \gamma_4 \left[\left\| \frac{\partial^{3m} \psi_1(x)}{\partial x_1^3 \partial x_2^3 \ldots \partial x_m^3} \right\|_{L_2(\Omega_l^m)} + \left\| \frac{\partial^{3m} \psi_2(x)}{\partial x_1^3 \partial x_2^3 \ldots \partial x_m^3} \right\|_{L_2(\Omega_l^m)} + \left\| \frac{\partial^{3m} \varphi_1(x)}{\partial x_1^3 \partial x_2^3 \ldots \partial x_m^3} \right\|_{L_2(\Omega_l^m)} \right.$$
$$\left. + \left\| \frac{\partial^{3m} \varphi_2(x)}{\partial x_1^3 \partial x_2^3 \ldots \partial x_m^3} \right\|_{L_2(\Omega_l^m)} + \left\| \frac{\partial^{3m} f_1(x, V_1)}{\partial x_1^3 \partial x_2^3 \ldots \partial x_m^3} \right\|_{L_2(\Omega_l^m)} \right] < \infty,$$

$$|g_2(x, \varepsilon, \omega, \nu)| \leq \sum_{n_1, \ldots, n_m = 1}^{\infty} |\vartheta_{n_1, \ldots, n_m}(x)| \left[|c_{2\, n_1, \ldots, n_m}(\varepsilon, \omega, \nu)| \right.$$
$$\left. + \left| \int_{\Omega_l^m} f_2 \left(y, \int_{\Omega_l^m} \Theta_2(z) \sum_{n_1, \ldots, n_m = 1}^{\infty} g_{2\, n_1, \ldots, n_m}(\varepsilon, \omega, \nu) \vartheta_{n_1, \ldots, n_m}(z)\, dz \right) \vartheta_{n_1, \ldots, n_m}(y)\, dy \right| \right]$$
$$\leq \gamma_4 \left[\left\| \frac{\partial^{3m} \psi_2(x)}{\partial x_1^3 \partial x_2^3 \ldots \partial x_m^3} \right\|_{L_2(\Omega_l^m)} + \left\| \frac{\partial^{3m} \varphi_1(x)}{\partial x_1^3 \partial x_2^3 \ldots \partial x_m^3} \right\|_{L_2(\Omega_l^m)} \right. \tag{59}$$
$$\left. + \left\| \frac{\partial^{3m} \varphi_2(x)}{\partial x_1^3 \partial x_2^3 \ldots \partial x_m^3} \right\|_{L_2(\Omega_l^m)} + \left\| \frac{\partial^{3m} f_2(x, V_2)}{\partial x_1^3 \partial x_2^3 \ldots \partial x_m^3} \right\|_{L_2(\Omega_l^m)} \right] < \infty,$$

where

$$\left\| \frac{\partial^{3m} f_i(x, V_i)}{\partial x_1^3 \partial x_2^3 \ldots \partial x_m^3} \right\|_{L_2(\Omega_l^m)} = \sqrt{\int_{\Omega_l^m} \left[\frac{\partial^{3m} f_i(x, V_i)}{\partial x_1^3 \partial x_2^3 \ldots \partial x_m^3} \right]^2 dx},$$

$$V_i = \int_{\Omega_l^m} \Theta_i(y) \sum_{n_1, \ldots, n_m = 1}^{\infty} g_{i\, n_1, \ldots, n_m}(\varepsilon, \omega, \nu) \vartheta_{n_1, \ldots, n_m}(z)\, dz, \quad i = 1, 2,$$

$$\gamma_4 = C_{03} C_{05} \left(\frac{2}{l} \right)^{\frac{3m}{2}} \left(\frac{l}{\pi} \right)^{3m}, \quad C_{05} = \max \{ C_{01};\ C_{02};\ 1 \}.$$

From (58) and (59) the convergence of series (57) is implied. Lemma 2 is proved. □

So, we determined the redefinition functions as a Fourier series (57). Using representations (39) and (40), we can present Fourier series (33) and (34) of the main unknown functions as

$$U(t, x, \varepsilon, \omega, \nu) = \sum_{n_1,\ldots,n_m=1}^{\infty} \vartheta_{n_1,\ldots,n_m}(x) \left[\psi_{1 n_1,\ldots,n_m} W_{11 n_1,\ldots,n_m}(t, \varepsilon, \omega, \nu) + \psi_{2 n_1,\ldots,n_m} W_{12 n_1,\ldots,n_m}(t, \varepsilon, \omega, \nu) \right.$$
$$\left. + \varphi_{1 n_1,\ldots,n_m} W_{13 n_1,\ldots,n_m}(t, \varepsilon, \omega, \nu) + \varphi_{2 n_1,\ldots,n_m} W_{14 n_1,\ldots,n_m}(t, \varepsilon, \omega, \nu) \right], \quad t > 0, \quad (60)$$

$$U(t, x, \varepsilon, \omega, \nu) = \sum_{n_1,\ldots,n_m=1}^{\infty} \vartheta_{n_1,\ldots,n_m}(x) \left[\psi_{2 n_1,\ldots,n_m} W_{21 n_1,\ldots,n_m}(t, \varepsilon, \omega, \nu) \right.$$
$$\left. + \varphi_{1 n_1,\ldots,n_m} W_{22 n_1,\ldots,n_m}(t, \varepsilon, \omega, \nu) + \varphi_{2 n_1,\ldots,n_m} W_{23 n_1,\ldots,n_m}(t, \varepsilon, \omega, \nu) \right], \quad t < 0, \quad (61)$$

where

$$W_{i1\, n_1,\ldots,n_m}(t, \varepsilon, \omega, \nu) = \Delta_{i1\, n_1,\ldots,n_m}(\varepsilon, \omega, \nu) Q_{i3 n_1,\ldots,n_m}(t, \varepsilon, \omega, \nu), \quad i = 1, 2,$$

$$W_{12 n_1,\ldots,n_m}(t, \varepsilon, \omega, \nu) = \Delta_{12 n_1,\ldots,n_m}(\varepsilon, \omega, \nu) Q_{13 n_1,\ldots,n_m}(t, \varepsilon, \omega, \nu) + \Delta_{21 n_1,\ldots,n_m}(\varepsilon, \omega, \nu) Q_{14 n_1,\ldots,n_m}(t, \varepsilon, \omega, \nu),$$

$$W_{1j n_1,\ldots,n_m}(t, \varepsilon, \omega, \nu) = Q_{1j-2\, n_1,\ldots,n_m}(t, \varepsilon, \omega, \nu) + \Delta_{1j n_1,\ldots,n_m}(\varepsilon, \omega, \nu) Q_{13\, n_1,\ldots,n_m}(t, \varepsilon, \omega, \nu)$$
$$+ \Delta_{2j-1\, n_1,\ldots,n_m}(\varepsilon, \omega, \nu) Q_{14\, n_1,\ldots,n_m}(t, \varepsilon, \omega, \nu), \quad j = 3, 4,$$

$$W_{2k n_1,\ldots,n_m}(t, \varepsilon, \omega, \nu) = Q_{2k-1 n_1,\ldots,n_m}(t, \varepsilon, \omega, \nu) + \Delta_{2k n_1,\ldots,n_m}(\varepsilon, \omega, \nu) Q_{23 n_1,\ldots,n_m}(t, \varepsilon, \omega, \nu), \quad k = 2, 3.$$

To establish the uniqueness of the function $U(t, x, \varepsilon, \omega, \nu)$, we suppose that there are two solutions U_1 and U_2 to this problem. Then, their difference $U = U_1 - U_2$ is a solution of Equation (1), satisfying conditions (2)–(6) with functions $\varphi_i(x) \equiv 0$, $\psi_i(x) \equiv 0$ ($i = 1, 2$). Then, for $\varphi_{i r_1,\ldots,n_m} = \psi_{i n_1,\ldots,n_m} = 0$ ($i = 1, 2$), it follows from Formulas (60) and (61) in the domain Ω that

$$\int_{\Omega_l^m} U(t, x, \varepsilon, \omega, \nu) \vartheta_{n_1,\ldots,n_m}(x) \, dx = 0.$$

Hence, by virtue of the completeness of the systems of eigenfunctions $\left\{ \sqrt{\frac{2}{l}} \sin \frac{\pi n_1}{l} x_1 \right\}$, $\left\{ \sqrt{\frac{2}{l}} \sin \frac{\pi n_2}{l} x_2 \right\}, \ldots, \left\{ \sqrt{\frac{2}{l}} \sin \frac{\pi n_m}{l} x_m \right\}$ in $L_2(\Omega_l^m)$, we deduce that $U(t, x, \varepsilon, \omega, \nu) \equiv 0$ for all $x \in \Omega_l^m \equiv [0; l]^m$ and $t \in [-T; T]$.

Therefore, for regular values of spectral parameters ω and ν, the function $U(t, x, \varepsilon, \omega, \nu)$ is a unique solution tp the mixed type integro-differential Equation (1) with conditions (2)–(6), if this function exists in the domain Ω.

Lemma 3. *Let smoothness conditions hold. Then, for regular values of spectral parameters ω and ν, series (60) and (61) converge. At the same time, their term by term differentiation is possible.*

Proof. According to the properties of the Mittag–Leffler function (Formulas (47) and (48)), the functions $W_{1 i n_1,\ldots,n_m}(t, \varepsilon, \omega, \nu)$ ($i = \overline{1, 4}$) and $W_{2 j n_1,\ldots,n_m}(t, \varepsilon, \omega, \nu)$ ($j = \overline{1, 3}$) are uniformly bounded on the segment $[-T; T]$. So, for any positive integers $n_1, \ldots, n_m$, there exist finite constant numbers C_{1k} ($k = 1, 2$); then, the following estimates take place

$$\max_{n_1,\ldots,n_m \in \mathbb{N}} \max_{i=\overline{1,4}} \left| W_{1 i n_1,\ldots,n_m}(t, \varepsilon, \omega, \nu) \right| \leq C_{11}, \quad \max_{n_1,\ldots,n_m \in \mathbb{N}} \max_{j=\overline{1,3}} \left| W_{1 j n_1,\ldots,n_m}(t, \varepsilon, \omega, \nu) \right| \leq C_{12}, \quad (62)$$

where $C_{1k} = \text{const}, \ k = 1, 2.$

Analogously to the estimates (58) and (59), by applying estimates (62), the Cauchy–Schwartz inequality and Bessel inequality for series (60) and (61), we

$$|U(t, x, \varepsilon, \omega, \nu)| \leq \sum_{n_1, \ldots, n_m=1}^{\infty} |u_{n_1, \ldots, n_m}^+(t, \varepsilon, \omega, \nu)| \cdot |\vartheta_{n_1, \ldots, n_m}(x)|$$

$$\leq \left(\sqrt{\frac{2}{l}}\right)^m C_{11} \sum_{n_1, \ldots, n_m=1}^{\infty} [|\psi_{1\,n_1, \ldots, n_m}| + |\psi_{2\,n_1, \ldots, n_m}| + |\varphi_{1\,n_1, \ldots, n_m}| + |\varphi_{2\,n_1, \ldots, n_m}|]$$

$$\leq \gamma_5 \left[\sqrt{\int_{\Omega_l^m} \left[\frac{\partial^{3m}\psi_1(x)}{\partial x_1^3 \partial x_2^3 \ldots \partial x_m^3}\right]^2 dx} + \sqrt{\int_{\Omega_l^m} \left[\frac{\partial^{3m}\psi_2(x)}{\partial x_1^3 \partial x_2^3 \ldots \partial x_m^3}\right]^2 dx}\right. \tag{63}$$

$$\left. + \sqrt{\int_{\Omega_l^m} \left[\frac{\partial^{3m}\varphi_1(x)}{\partial x_1^3 \partial x_2^3 \ldots \partial x_m^3}\right]^2 dx} + \sqrt{\int_{\Omega_l^m} \left[\frac{\partial^{3m}\varphi_2(x)}{\partial x_1^3 \partial x_2^3 \ldots \partial x_m^3}\right]^2 dx}\right] < \infty,$$

where $\gamma_5 = \left(\sqrt{\frac{2}{l}}\right)^{\frac{3m}{2}} C_{11} C_{03} \left(\frac{l}{\pi}\right)^{3m}$,

$$|U(t, x, \varepsilon, \omega, \nu)| \leq \sum_{n_1, \ldots, n_m=1}^{\infty} |u_{n_1, \ldots, n_m}^-(t, \varepsilon, \omega, \nu)| \cdot |\vartheta_{n_1, \ldots, n_m}(x)|$$

$$\leq \gamma_6 \left[\sqrt{\int_{\Omega_l^m} \left[\frac{\partial^{3m}\psi_2(x)}{\partial x_1^3 \partial x_2^3 \ldots \partial x_m^3}\right]^2 dx}\right. \tag{64}$$

$$\left. + \sqrt{\int_{\Omega_l^m} \left[\frac{\partial^{3m}\varphi_1(x)}{\partial x_1^3 \partial x_2^3 \ldots \partial x_m^3}\right]^2 dx} + \sqrt{\int_{\Omega_l^m} \left[\frac{\partial^{3m}\varphi_2(x)}{\partial x_1^3 \partial x_2^3 \ldots \partial x_m^3}\right]^2 dx}\right] < \infty,$$

where $\gamma_6 = \left(\sqrt{\frac{2}{l}}\right)^{\frac{3m}{2}} C_{12} C_{03} \left(\frac{l}{\pi}\right)^{3m}$.

It follows from estimates (63) and (64) that the series (60) and (61) are convergent absolutely and uniformly in the domain $\overline{\Omega}$ for the

$$(n_1, \ldots, n_m, \omega, \nu) \in \aleph = \{n_1, \ldots, n_m \in \mathbb{N}; \ \omega \in \Lambda_1; \ \nu \in \Lambda_2\}.$$

Therefore, for the $(n_1, \ldots, n_m, \omega, \nu) \in \aleph$ functions, (63) and (64) formally differentiate in $\overline{\Omega}$ the required number of times

$$_cD_{0t}^{\alpha_1} U(t, x, \varepsilon, \omega, \nu) = \sum_{n_1, \ldots, n_m=1}^{\infty} \vartheta_{n_1, \ldots, n_m}(x)$$

$$\times [\psi_{1\,n_1, \ldots, n_m}\,{}_cD_{0t}^{\alpha_1} W_{11\,n_1, \ldots, n_m}(t, \varepsilon, \omega, \nu) + \psi_{2\,n_1, \ldots, n_m}\,{}_cD_{0t}^{\alpha_1} W_{12\,n_1, \ldots, n_m}(t, \varepsilon, \omega, \nu) \tag{65}$$

$$+ \varphi_{1\,n_1, \ldots, n_m}\,{}_cD_{0t}^{\alpha_1} W_{13\,n_1, \ldots, n_m}(t, \varepsilon, \omega, \nu) + \varphi_{2\,n_1, \ldots, n_m}\,{}_cD_{0t}^{\alpha_1} W_{14\,n_1, \ldots, n_m}(t, \varepsilon, \omega, \nu)], \ t > 0,$$

$$_cD_{0t}^{\alpha_2} U(t, x, \varepsilon, \omega, \nu) = \sum_{n_1, \ldots, n_m=1}^{\infty} \vartheta_{n_1, \ldots, n_m}(x) [\psi_{2\,n_1, \ldots, n_m}\,{}_cD_{0t}^{\alpha_2} W_{21\,n_1, \ldots, n_m}(t, \varepsilon, \omega, \nu)$$

$$+ \varphi_{1\,n_1, \ldots, n_m}\,{}_cD_{0t}^{\alpha_2} W_{22\,n_1, \ldots, n_m}(t, \varepsilon, \omega, \nu) + \varphi_{2\,n_1, \ldots, n_m}\,{}_cD_{0t}^{\alpha_2} W_{23\,n_1, \ldots, n_m}(t, \varepsilon, \omega, \nu)], \ t < 0, \tag{66}$$

$$U_{x_1 x_1}(t, x, \varepsilon, \omega, \nu) = -\sum_{n_1, \ldots, n_m=1}^{\infty} \left(\frac{\pi n_1}{l}\right)^2 \vartheta_{n_1, \ldots, n_m}(x) [\psi_{1\,n_1, \ldots, n_m} W_{11\,n_1, \ldots, n_m}(t, \varepsilon, \omega, \nu)$$

$$+ \psi_{2\,n_1, \ldots, n_m} W_{12\,n_1, \ldots, n_m}(t, \varepsilon, \omega, \nu) + \varphi_{1\,n_1, \ldots, n_m} W_{13\,n_1, \ldots, n_m}(t, \varepsilon, \omega, \nu) \tag{67}$$

$$+ \varphi_{2\,n_1, \ldots, n_m} W_{14\,n_1, \ldots, n_m}(t, \varepsilon, \omega, \nu)], \ t > 0,$$

$$U_{x_1 x_1}(t, x, \varepsilon, \omega, \nu) = -\sum_{n_1, \ldots, n_m=1}^{\infty} \left(\frac{\pi n_1}{l}\right)^2 \vartheta_{n_1, \ldots, n_m}(x) [\psi_{2\,n_1, \ldots, n_m} W_{21\,n_1, \ldots, n_m}(t, \varepsilon, \omega, \nu)$$

$$+ \varphi_{1\,n_1, \ldots, n_m} W_{22\,n_1, \ldots, n_m}(t, \varepsilon, \omega, \nu) + \varphi_{2\,n_1, \ldots, n_m} W_{23\,n_1, \ldots, n_m}(t, \varepsilon, \omega, \nu)], \ t < 0, \tag{68}$$

$$U_{x_2x_2}(t, x, \varepsilon, \omega, \nu) = -\sum_{n_1,\ldots,n_m=1}^{\infty} \left(\frac{\pi n_2}{l}\right)^2 \vartheta_{n_1,\ldots,n_m}(x) \left[\psi_{1\,n_1,\ldots,n_m} W_{11\,n_1,\ldots,n_m}(t, \varepsilon, \omega, \nu)\right.$$
$$+\psi_{2\,n_1,\ldots,n_m} W_{12\,n_1,\ldots,n_m}(t, \varepsilon, \omega, \nu) + \varphi_{1\,n_1,\ldots,n_m} W_{13\,n_1,\ldots,n_m}(t, \varepsilon, \omega, \nu) \quad (69)$$
$$\left.+\varphi_{2\,n_1,\ldots,n_m} W_{14\,n_1,\ldots,n_m}(t, \varepsilon, \omega, \nu)\right], \quad t > 0,$$

$$U_{x_2x_2}(t, x, \varepsilon, \omega, \nu) = -\sum_{n_1,\ldots,n_m=1}^{\infty} \left(\frac{\pi n_2}{l}\right)^2 \vartheta_{n_1,\ldots,n_m}(x) \left[\psi_{2\,n_1,\ldots,n_m} W_{21\,n_1,\ldots,n_m}(t, \varepsilon, \omega, \nu)\right.$$
$$\left.+ \varphi_{1\,n_1,\ldots,n_m} W_{22\,n_1,\ldots,n_m}(t, \varepsilon, \omega, \nu) + \varphi_{2\,n_1,\ldots,n_m} W_{23\,n_1,\ldots,n_m}(t, \varepsilon, \omega, \nu)\right], \quad t < 0. \quad (70)$$

The expansions of the following functions into Fourier series are defined in the domain Ω in a similar way

$$U_{x_3x_3}(t, x, \varepsilon, \omega, \nu), \ldots, U_{x_mx_m}(t, x, \varepsilon, \omega, \nu), \ _CD_{0t}^{\alpha_1}U_{x_1x_1}(t, x, \varepsilon, \omega, \nu),$$

$$_CD_{0t}^{\alpha_2}U_{x_1x_1}(t, x, \varepsilon, \omega, \nu), \ _CD_{0t}^{\alpha_1}U_{x_2x_2}(t, x, \varepsilon, \omega, \nu), \ldots, \ _CD_{0t}^{\alpha_2}U_{x_2x_2}(t, x, \varepsilon, \omega, \nu), \ldots,$$

$$_CD_{0t}^{\alpha_1}U_{x_mx_m}(t, x, \varepsilon, \omega, \nu), \ _CD_{0t}^{\alpha_2}U_{x_mx_m}(t, x, \varepsilon, \omega, \nu).$$

The convergence of series (65) and (66) is proved similarly to the proof of the convergence of series (60) and (61). So, it is enough to show the convergence of series (67) and (70). Taking into account Formulas (42)–(44) and estimates (62) and applying the Cauchy–Schwartz inequality and Bessel inequality, we obtain

$$|U_{x_1x_1}(t, x, \varepsilon, \omega, \nu)| \le \sum_{n_1,\ldots,n_m=1}^{\infty} \left(\frac{\pi n_1}{l}\right)^2 |u_{n_1,\ldots,n_m}^+(t, \varepsilon, \omega, \nu)| \cdot |\vartheta_{n_1,\ldots,n_m}(x)|$$

$$\le \left(\sqrt{\frac{2}{l}}\right)^m \left(\frac{\pi}{l}\right)^2 C_{11} \sum_{n_1,\ldots,n_m=1}^{\infty} n_1^2 [|\psi_{1\,n_1,\ldots,n_m}| + |\psi_{2\,n_1,\ldots,n_m}| + |\varphi_{1\,n_1,\ldots,n_m}| + |\varphi_{2\,n_1,\ldots,n_m}|]$$

$$\le \left(\sqrt{\frac{2}{l}}\right)^m C_{11} \left(\frac{l}{\pi}\right)^{3m-2} \left[\sum_{n_1,\ldots,n_m=1}^{\infty} \frac{1}{n_1 n_2^3 \ldots n_m^3} |\psi_{1\,n_1,\ldots,n_m}^{(3m)}| + \sum_{n_1,\ldots,n_m=1}^{\infty} \frac{1}{n_1 n_2^3 \ldots n_m^3} |\psi_{2\,n_1,\ldots,n_m}^{(3m)}|\right.$$
$$\left.+ \sum_{n_1,\ldots,n_m=1}^{\infty} \frac{1}{n_1 n_2^3 \ldots n_m^3} |\varphi_{1\,n_1,\ldots,n_m}^{(3m)}| + \sum_{n_1,\ldots,n_m=1}^{\infty} \frac{1}{n_1 n_2^3 \ldots n_m^3} |\varphi_{2\,n_1,\ldots,n_m}^{(3m)}|\right]$$

$$\le \gamma_7 \left[\sqrt{\int_{\Omega_l^m} \left[\frac{\partial^{3m}\psi_1(x)}{\partial x_1^3 \partial x_2^3 \ldots \partial x_m^3}\right]^2 dx} + \sqrt{\int_{\Omega_l^m} \left[\frac{\partial^{3m}\psi_2(x)}{\partial x_1^3 \partial x_2^3 \ldots \partial x_m^3}\right]^2 dx}\right.$$
$$\left.+ \sqrt{\int_{\Omega_l^m} \left[\frac{\partial^{3m}\varphi_1(x)}{\partial x_1^3 \partial x_2^3 \ldots \partial x_m^3}\right]^2 dx} + \sqrt{\int_{\Omega_l^m} \left[\frac{\partial^{3m}\varphi_2(x)}{\partial x_1^3 \partial x_2^3 \ldots \partial x_m^3}\right]^2 dx}\right] < \infty,$$

where $\gamma_7 = \left(\sqrt{\frac{2}{l}}\right)^{\frac{3m}{2}} C_{11} C_{06} \left(\frac{l}{\pi}\right)^{3m-2}$, $C_{06} = \sqrt{\sum_{n_1,\ldots,n_m=1}^{\infty} \frac{1}{n_1^2 n_2^6 \ldots n_m^6}}$;

$$|U_{x_2x_2}(t, x, \varepsilon, \omega, \nu)| \le \sum_{n_1,\ldots,n_m=1}^{\infty} \left(\frac{\pi n_2}{l}\right)^2 |u_{n_1,\ldots,n_m}^-(t, \varepsilon, \omega, \nu)| \cdot |\vartheta_{n_1,\ldots,n_m}(x)|$$

$$\le \left(\sqrt{\frac{2}{l}}\right)^m \left(\frac{\pi}{l}\right)^2 C_{12} \sum_{n_1,\ldots,n_m=1}^{\infty} n_2^2 [|\psi_{2\,n_1,\ldots,n_m}| + |\varphi_{1\,n_1,\ldots,n_m}| + |\varphi_{2\,n_1,\ldots,n_m}|]$$

$$\leq \left(\sqrt{\frac{2}{l}}\right)^m C_{12} \left(\frac{1}{\pi}\right)^{3m-2} \left[\sum_{n_1,\ldots,n_m=1}^{\infty} \frac{1}{n_1^3 n_2^3 n_3^3 \ldots n_m^3} \left|\psi_{2n_1,\ldots,n_m}^{(3m)}\right|\right.$$

$$+ \sum_{n_1,\ldots,n_m=1}^{\infty} \frac{1}{n_1^3 n_2^3 n_3^3 \ldots n_m^3} \left|\varphi_{1n_1,\ldots,n_m}^{(3m)}\right| + \sum_{n_1,\ldots,n_m=1}^{\infty} \frac{1}{n_1^3 n_2^3 n_3^3 \ldots n_m^3} \left|\varphi_{2n_1,\ldots,n_m}^{(3m)}\right|\right]$$

$$\leq \gamma_8 \left[\sqrt{\int_{\Omega_l^m} \left[\frac{\partial^{3m} \psi_2(x)}{\partial x_1^3 \partial x_2^3 \ldots \partial x_m^3}\right]^2 dx}\right.$$

$$+ \sqrt{\int_{\Omega_l^m} \left[\frac{\partial^{3m} \varphi_1(x)}{\partial x_1^3 \partial x_2^3 \ldots \partial x_m^3}\right]^2 dx} + \sqrt{\int_{\Omega_l^m} \left[\frac{\partial^{3m} \varphi_2(x)}{\partial x_1^3 \partial x_2^3 \ldots \partial x_m^3}\right]^2 dx}\right] < \infty,$$

where $\gamma_8 = \left(\sqrt{\frac{2}{l}}\right)^{\frac{3m}{2}} C_{12} C_{07} \left(\frac{1}{\pi}\right)^{3m-2}$, $C_{07} = \sqrt{\sum_{n_1,\ldots,n_m=1}^{\infty} \frac{1}{n_1^6 n_2^6 \ldots n_m^6}}$.

The convergence of series (68) and (69) is similar to the convergence of series (67) and (70). The convergence of Fourier series for functions

$$U_{x_3 x_3}(t, x, \varepsilon, \omega, \nu), \ldots, U_{x_m x_m}(t, x, \varepsilon, \omega, \nu), \, _C D_{0t}^{\alpha_1} U_{x_1 x_1}(t, x, \varepsilon, \omega, \nu),$$

$$_C D_{0t}^{\alpha_2} U_{x_1 x_1}(t, x, \varepsilon, \omega, \nu), \, _C D_{0t}^{\alpha_1} U_{x_2 x_2}(t, x, \varepsilon, \omega, \nu), \ldots, \, _C D_{0t}^{\alpha_2} U_{x_2 x_2}(t, x, \varepsilon, \omega, \nu), \ldots,$$

$$_C D_{0t}^{\alpha_1} U_{x_m x_m}(t, x, \varepsilon, \omega, \nu), \, _C D_{0t}^{\alpha_2} U_{x_m x_m}(t, x, \varepsilon, \omega, \nu)$$

is proved in a similar way in the domain Ω. It follows from these last estimates that functions (60) and (61) possess the properties of (2) for the regular values of spectral parameters ω and ν. □

7. Continuous Dependence of Solution on the Small Parameter

We consider the continuous dependence of the solution to the problem (1)–(4) on small-parameter $\varepsilon > 0$ for regular values of spectral parameters ω and ν. Let ε_1 and ε_2 be two different values of small positive parameter ε. It is easy to check from (47)–(49) that the following estimates hold

$$\max_{n_1,\ldots,n_m \in \mathbb{N}} \max_{t \in [0;T]} \left|W_{1in_1,\ldots,n_m}(t, \varepsilon_1, \omega, \nu) - W_{1in_1,\ldots,n_m}(t, \varepsilon_2, \omega, \nu)\right| \leq C_{21} |\varepsilon_1 - \varepsilon_2|, \, i = \overline{1,4}, \quad (71)$$

$$\max_{n_1,\ldots,n_m \in \mathbb{N}} \max_{t \in [-T;0]} \left|W_{2in_1,\ldots,n_m}(t, \varepsilon_1, \omega, \nu) - W_{2in_1,\ldots,n_m}(t, \varepsilon_2, \omega, \nu)\right| \leq C_{22} |\varepsilon_1 - \varepsilon_2|, \, i = \overline{1,3}, \quad (72)$$

where $0 < C_{2i} = \text{const}$, $\varepsilon_i \in (0; \varepsilon_0)$, $0 < \varepsilon_0 = \text{const}$, $i = 1, 2$.

Then, taking estimates (63), (64), (71) and (72) into account and applying the Cauchy–Schwartz inequality and Bessel inequality, from series (60) and (61), we obtain

$$|U(t, x, \varepsilon_1, \omega, \nu) - U(t, x, \varepsilon_2, \omega, \nu)|$$

$$\leq \sum_{n_1,\ldots,n_m=1}^{\infty} \left|u_{n_1,\ldots,n_m}^+(t, \varepsilon_1, \omega, \nu) - u_{n_1,\ldots,n_m}^+(t, \varepsilon_2, \omega, \nu)\right| \cdot \left|\vartheta_{n_1,\ldots,n_m}(x)\right|$$

$$\leq \left(\sqrt{\frac{2}{l}}\right)^m C_{21} |\varepsilon_1 - \varepsilon_2| \sum_{n_1,\ldots,n_m=1}^{\infty} \left[|\psi_{1n_1,\ldots,n_m}| + |\psi_{2n_1,\ldots,n_m}| + |\varphi_{1n_1,\ldots,n_m}| + |\varphi_{2n_1,\ldots,n_m}|\right] \quad (73)$$

$$\leq \gamma_9 |\varepsilon_1 - \varepsilon_2| \left[\sqrt{\int_{\Omega_l^m} \left[\frac{\partial^{3m} \psi_1(x)}{\partial x_1^3 \partial x_2^3 \ldots \partial x_m^3}\right]^2 dx} + \sqrt{\int_{\Omega_l^m} \left[\frac{\partial^{3m} \psi_2(x)}{\partial x_1^3 \partial x_2^3 \ldots \partial x_m^3}\right]^2 dx}\right.$$

$$\left. + \sqrt{\int_{\Omega_l^m} \left[\frac{\partial^{3m} \varphi_1(x)}{\partial x_1^3 \partial x_2^3 \ldots \partial x_m^3}\right]^2 dx} + \sqrt{\int_{\Omega_l^m} \left[\frac{\partial^{3m} \varphi_2(x)}{\partial x_1^3 \partial x_2^3 \ldots \partial x_m^3}\right]^2 dx}\right] = |\varepsilon_1 - \varepsilon_2| \cdot C_{31},$$

where $\gamma_9 = \left(\sqrt{\frac{2}{l}}\right)^{\frac{3m}{2}} C_{21} C_{03} \left(\frac{l}{\pi}\right)^{3m}, 0 < C_{31} = \text{const} < \infty;$

$$|U(t, x, \varepsilon_1, \omega, v) - U(t, x, \varepsilon_2, \omega, v)|$$
$$\leq \sum_{n_1, \ldots, n_m = 1}^{\infty} |u_{n_1, \ldots, n_m}(t, \varepsilon_1, \omega, v) - u_{n_1, \ldots, n_m}(t, \varepsilon_2, \omega, v)| \cdot |\vartheta_{n_1, \ldots, n_m}(x)|$$
$$\leq \gamma_{10} |\varepsilon_1 - \varepsilon_2| \left[\sqrt{\int_{\Omega_l^m} \left[\frac{\partial^{3m} \psi_2(x)}{\partial x_1^3 \partial x_2^3 \ldots \partial x_m^3}\right]^2 dx} + \right. \tag{74}$$
$$\left. + \sqrt{\int_{\Omega_l^m} \left[\frac{\partial^{3m} \varphi_1(x)}{\partial x_1^3 \partial x_2^3 \ldots \partial x_m^3}\right]^2 dx} + \sqrt{\int_{\Omega_l^m} \left[\frac{\partial^{3m} \varphi_2(x)}{\partial x_1^3 \partial x_2^3 \ldots \partial x_m^3}\right]^2 dx} \right] = |\varepsilon_1 - \varepsilon_2| \cdot C_{32},$$

where $\gamma_{10} = \left(\sqrt{\frac{2}{l}}\right)^{\frac{3m}{2}} C_{22} C_{03} \left(\frac{l}{\pi}\right)^{3m}, 0 < C_{32} = \text{const} < \infty.$

It follows from estimates (73) and (74) that $|U(t, x, \varepsilon_1, \omega, v) - U(t, x, \varepsilon_2, \omega, v)|$ is small if $|\varepsilon_1 - \varepsilon_2|$ is small in the domain $\overline{\Omega}$ for the $(n_1, \ldots, n_m, \omega, v) \in \aleph$.

8. Conclusions and Statement of the Theorem

In the present paper, we study the questions of the one-value solvability of an inverse boundary value problem (1)–(6) for a mixed type integro-differential equation with Caputo operators of different fractional orders and spectral parameters in a multidimensional rectangular domain. For $(n_1, \ldots, n_m, \omega, v) \in \aleph$, we proved four lemmas under the following conditions **A**: Let functions

$$\varphi_i(x), \psi_i(x) \in C^2(\Omega_l^m), \quad f_i\left(x, \int_{\Omega_l^m} \Theta_i(y) g_i(y) \, dy\right) \in C_x^2(\Omega_l^m \times \mathbb{R}), \quad i = 1, 2$$

in the domain Ω_l^m have piecewise continuous third order derivatives.

We will formulate a theorem as generalizing the above four proved lemmas. Thus, the following theorem is true.

Theorem 1. *Let the conditions of A be fulfilled. Then, for the possible numbers $n_1, \ldots, n_m$ and regular values of spectral parameters ω and v from the set $\aleph$, the inverse boundary value problem (1)–(6) is uniquely solvable in the domain Ω and the triple of solutions is represented in the form of series (57), (60) and (61). Moreover, it is true that*

$$\lim_{\varepsilon \to 0} U(t, x, \varepsilon, \omega, v) = U(t, x, 0, \omega, v),$$

where $U(t, x, 0, \omega, v)$ is the solution of the mixed type fractional integro-differential equation of the form

$$A_0(U) - B_\omega(U) = \begin{cases} v \int_0^T K_1(t, s) U(s, x) \, ds + F_1(t, x), & t > 0, \\ v \int_{-T}^0 K_2(t, s) U(s, x) \, ds + F_2(t, x), & t < 0, \end{cases}$$

$$A_0(U) = \left[\frac{1 + \text{sgn}(t)}{2} {}_cD_{0t}^{\alpha_1} + \frac{1 - \text{sgn}(t)}{2} {}_cD_{0t}^{\alpha_2}\right] U(t, x), \quad B_\omega(U) = \begin{cases} \sum_{i=1}^m U_{x_i x_i}, & t > 0, \\ \omega^2 \sum_{i=1}^m U_{x_i x_i}, & t < 0 \end{cases}$$

with boundary value conditions (3)–(6) under consideration,

$$F_i(t,x) = k_i(t)\left[g_i(x) + f_i\left(x, \int_{\Omega_l^m} \Theta_i(y)\, g_i(y)\, dy\right)\right], \quad i = 1, 2.$$

As a conclusion, we say that the numerical methods for solving fractional differential equations are important in the implementation of applied problems. In the future, we will also try to consider the applications of the numerical solution to the problems that we are solving. There are many methods for the numerical implementation of fractional differential equations. In this regard, we note the work done in [42]. In this paper, a new class of (C, G_f)-invex functions is introduced and given nontrivial numerical examples, which justify the existence of such functions. Moreover, we construct generalized convexity definitions (such as, (F, G_f)-invexity, C-convex etc.).

Author Contributions: Conceptualization, T.K.Y. and E.T.K. All authors have read and agreed to the published version of the manuscript.

Funding: This research received no external funding.

Conflicts of Interest: The authors declare no conflicts of interest.

References

1. Samko, S.G.; Kilbas, A.A.; Marichev, O.I. *Fractional Integrals and Derivatives. Theory and Applications*; Gordon and Breach: Yverdon, Switzerland, 1993.
2. Mainardi, F. Fractional calculus: some basic problems in continuum and statistical mechanics. In *Fractals and Fractional Calculus in Continuum Mechanics*; Carpinteri A., Mainardi F., Eds.; Springer: Wien, Austria, 1997.
3. Area, I.; Batarfi, H.; Losada, J.; Nieto, J.J.; Shammakh, W.; Torres, A. On a fractional order Ebola epidemic model. *Adv. Differ. Equ.* **2015**, 278, doi:10.1186/s13662-015-0613-5. [CrossRef]
4. Hussain, A.; Baleanu, D.; Adeel, M. Existence of solution and stability for the fractional order novel coronavirus (nCoV-2019) model. *Adv. Differ. Equ.* **2020**, 384. [CrossRef] [PubMed]
5. Ullah, S.; Khan, M.A.; Farooq, M.; Hammouch, Z.; Baleanu, D. A fractional model for the dynamics of tuberculosis infection using Caputo-Fabrizio derivative. *Discret. Contin. Dyn. Syst. Ser. S* **2020**, 13, 975–993. [CrossRef]
6. Tenreiro Machado, J.A. *Handbook of Fractional Calculus with Applications in 8 Volumes*; Walter de Gruyter GmbH: Berlin, Germany; Boston, MA, USA, 2019.
7. Kumar, D.; Baleanu, D. Fractional Calculus and Its Applications in Physics. *Front. Phys.* **2019**, 7, 81. [CrossRef]
8. Sun, H.; Chang, A.; Zhang, Y.; Chen, W. A review on variable-order fractional differential equations: Mathematical foundations, physical models, numerical methods and applications. *Fract. Calc. Appl. Anal.* **2019**, 22, 27–59. [CrossRef]
9. Saxena Ram, K.; Garra, R.; Orsingher, E. Analytical solution of space-time fractional telegraph-type equations involving Hilfer and Hadamard derivatives. *Integral Transform. Spec. Funct.* **2015**, 27, 30–42. [CrossRef]
10. Patnaik, S.; Hollkamp, J.P.; Semperlotti, F. Applications of variable-order fractional operators: A review. *Proc. R. Soc.* **2020**, 476, 2234. [CrossRef]
11. Garra, R.; Gorenflo, R.; Polito, F.; Tomovski, Ž. Hilfer-Prabhakar derivatives and some applications. *Appl. Math. Comput.* **2014**, 242, 576–589. [CrossRef]
12. Tenreiro Machado, J.A. (Ed.) *Handbook of Fractional Calculus with Applications*; Walter de Gruyter GmbH: Berlin, Germany, 2019; Volumes 8.
13. Hilfer, R. *Application of Fractional Calculus in Physics*; World Scientific Publishing Company: Singapore, 2000.
14. Hilfer, R. On fractional relaxation. *Fractals* **2003**, 11, 251–257. [CrossRef]
15. Hilfer, R. Experimental evidence for fractional time evolution in glass forming materials. *Chem. Phys.* **2002**, 284, 399–408. [CrossRef]
16. Klafter, J.; Lim, S.C.; Metzler, R. *Fractional Dynamics, Recent Advances*; World Scientific: Singapore, 2011; Chapter 9.

17. Sandev, T.; Tomovski, Z. *Fractional Equations and Models: Theory and Applications*; Springer Nature Switzerland AG: Cham, Switzerland, 2019.
18. Xu, C.; Yu, Y.; Chen, Y.Q.; Lu, Z. Forecast analysis of the epidemic trend of COVID-19 in the United States by a generalized fractional-order SEIR model. *medRxiv* **2020**. [CrossRef]
19. Cesarano, C. Generalized special functions in the description of fractional diffusive equations. *Commun. Appl. Ind. Math.* **2019**, *10*, 31–40. [CrossRef]
20. Assante, D.; Cesarano, C.; Fornaro, C.; Vazquez, L. Higher Order and Fractional Diffusive Equations. *J. Eng. Sci. Technol. Rev.* **2015**, *8*, 202–204. [CrossRef]
21. Dattoli, G.; Cesarano, C.; Ricci, P.; Vazquez, L. Special Polynomials and Fractional Calculus. *Math. Comput. Model.* **2003**, *37*, 729–733. [CrossRef]
22. Restrepo, J.; Ruzhansky, M.; Suragan, D. Explicit representations of solutions for linear fractional differential equations with variable coefficients. *arXiv* **2020**, arXiv:2006.1535v1.
23. Gel'fand, I.M. Some questions of analysis and differential equations. *Uspekhi Mat. Nauk.* **1959**, *14*, 3–19. (In Russian)
24. Uflyand, Y.S. On oscillation propagation in compound electric lines. *Inzhenerno-Phizicheskiy Zhurnal* **1964**, *7*, 89–92. (In Russian)
25. Terlyga, O.; Bellout, H.; Bloom, F. A hyperbolic-parabolic system arising in pulse combustion: existence of solutions for the linearized problem. *Electron. J. Differ. Equ.* **2013**, *2013*, 1–42.
26. Abdullaev, O.K.; Sadarangani, K. Nonlocal problems with integral gluing condition for loaded mixed type equations involving the Caputo fractional derivative. *Electron. J. Differ. Equ.* **2016**, *2016*, 164, 1–10. Available online: http://ejde.math.txstate.edu (accessed on 25 September 2020).
27. Agarwal, P.; Berdyshev, A.S.; Karimov, E.T. Solvability of a nonlocal problem with integral transmitting condition for mixed type equation with Caputo fractional derivative. *Results Math.* **2017**, *71*, 1235–1257. [CrossRef]
28. Zarubin, A.N. Boundary value problem for a differential-difference mixed-compound equation with fractional derivative and with functional delay and advance. *Differ. Equ.* **2019**, *55*, 220–230. [CrossRef]
29. Karimov, E.T.; Al-Salti, N.; Kerbal, S. An inverse source non-local problem for a mixed type equation with a Caputo fractional differential operator. *East Asian J. Appl. Math.* **2017**, *7*, 417–438. doi:10.4208/eajam.051216.280217aS2079736217000268. [CrossRef]
30. Karimov, E.T.; Kerbal, S.; Al-Salti, N. Inverse source problem for multi-term fractional mixed type equation. In *Advanes in Real and Complex Analysis with Applications*; Springer Nature Singapore Pte Ltd.: Singapore, 2017; pp. 289–301. [CrossRef]
31. Repin, O.A. Nonlocal problem with Saigo operators for mixed type equation of the third order. *Russ. Math.* **2019**, *63*, 55–60. [CrossRef]
32. Repin, O.A. On a problem for a mixed-type equation with fractional derivative. *Russ. Math.* **2018**, *62*, 38–42. [CrossRef]
33. Salakhitdinov, M.S.; Karimov, E.T. Uniqueness of an inverse source non-local problem for fractional order mixed type equations. *Eurasian Math. J.* **2016**, *7*, 74–83. Available online: http://mi.mathnet.ru/rus/emj/v7/i1/p74 (accessed on 25 September 2020).
34. Yuldashev, T.K.; Kadirkulov, B.J. Boundary value problem for weak nonlinear partial differential equations of mixed type with fractional Hilfer operator. *Axioms* **2020**, *9*, 68. [CrossRef]
35. Yuldashev, T.K.; Kadirkulov, B.J. Nonlocal problem for a mixed type fourth-order differential equation with Hilfer fractional operator. *Ural Math. J.* **2020**, *6*, 153–167. [CrossRef]
36. Yuldashev, T.K. Nonlocal boundary value problem for a nonlinear Fredholm integro-differential equation with degenerate kernel. *Differ. Equ.* **2018**, *54*, 1646–1653. [CrossRef]
37. Yuldashev, T.K. On the solvability of a boundary value problem for the ordinary Fredholm integrodifferential equation with a degenerate kernel. *Comput. Math. Math. Phys.* **2019**, *59*, 241–252. [CrossRef]
38. Yuldashev, T.K. Spectral features of the solving of a Fredholm homogeneous integro-differential equation with integral conditions and reflecting deviation. *Lobachevskii J. Math.* **2019**, *40*, 2116–2123. [CrossRef]
39. Yuldashev, T.K. On a boundary-value problem for Boussinesq type nonlinear integro-differential equation with reflecting argument. *Lobachevskii J. Math.* **2020**, *41*, 111–123. [CrossRef]

40. Yuldashev, T.K. On an integro-differential equation of pseudoparabolic-pseudohyperbolic type with degenerate kernels. *Proc. YSU Phys. Math. Sci.* **2018**, *52*, 19–26. Available online: http://mi.mathnet.ru/rus/uzeru/v52/i1/p1914 (accessed on 25 September 2020). [CrossRef]
41. Yuldashev, T.K. Nonlocal inverse problem for a pseudohyperbolic-pseudoelliptic type integro-differential equations. *Axioms* **2020**, *9*, 45, doi:10.3390/axioms9020045. [CrossRef]
42. Dubey R., Mishra L. N., Cesarano C. Multiobjective fractional symmetric duality in mathematical programming with (C, G_f)-invexity assumptions. *Axioms* **2019**, *8*, 97. [CrossRef]

© 2020 by the authors. Licensee MDPI, Basel, Switzerland. This article is an open access article distributed under the terms and conditions of the Creative Commons Attribution (CC BY) license (http://creativecommons.org/licenses/by/4.0/).

Article

Distributed-Order Non-Local Optimal Control

Faïçal Ndaïrou [†,‡] and Delfim F. M. Torres *,[‡]

Center for Research and Development in Mathematics and Applications (CIDMA), Department of Mathematics, University of Aveiro, 3810-193 Aveiro, Portugal; faical@ua.pt
* Correspondence: delfim@ua.pt; Tel.: +351-234-370-668
† This research is part of first author's Ph.D. project, which is carried out at the University of Aveiro under the Doctoral Program in Applied Mathematics of Universities of Minho, Aveiro, and Porto (MAP-PDMA).
‡ These authors contributed equally to this work.

Received: 9 September 2020; Accepted: 22 October; Published: 25 October 2020

Abstract: Distributed-order fractional non-local operators were introduced and studied by Caputo at the end of the 20th century. They generalize fractional order derivatives/integrals in the sense that such operators are defined by a weighted integral of different orders of differentiation over a certain range. The subject of distributed-order non-local derivatives is currently under strong development due to its applications in modeling some complex real world phenomena. Fractional optimal control theory deals with the optimization of a performance index functional, subject to a fractional control system. One of the most important results in classical and fractional optimal control is the Pontryagin Maximum Principle, which gives a necessary optimality condition that every solution to the optimization problem must verify. In our work, we extend the fractional optimal control theory by considering dynamical system constraints depending on distributed-order fractional derivatives. Precisely, we prove a weak version of Pontryagin's maximum principle and a sufficient optimality condition under appropriate convexity assumptions.

Keywords: distributed-order fractional calculus; basic optimal control problem; Pontryagin extremals

MSC: 26A33; 49K15

1. Introduction

Distributed-order fractional operators were introduced and studied by Caputo at the end of the previous century [1,2]. They can be seen as a kind of generalization of fractional order derivatives/integrals in the sense that these operators are defined by a weighted integral of different orders of differentiation over a certain range. This subject gained more interest at the beginning of the current century by researchers from different mathematical disciplines, through attempts to solve differential equations with distributed-order derivatives [3–6]. Moreover, at the same time, in the domain of applied mathematics, those distributed-order fractional operators have started to be used, in a satisfactory way, to describe some complex phenomena modeling real world problems—see, for instance, works in viscoelasticity [7,8] and in diffusion [9]. Today, the study of distributed-order systems with fractional derivatives is a hot subject—see, e.g., [10–12] and references therein.

Fractional optimal control deals with optimization problems involving fractional differential equations, as well as a performance index functional. One of the most important results is the Pontryagin Maximum Principle, which gives a first-order necessary optimality condition that every solution to the dynamic optimization problem must verify. By applying such a result, it is possible to find and identify candidate solutions to the optimal control problem. For the state of the art on fractional optimal control, we refer the readers to [13–15] and references therein. Recently, distributed-order fractional problems of the calculus of variations were introduced and investigated in [16]. Here, our main aim is to extend the distributed-order fractional Euler–Lagrange equation of [16] to the Pontryagin setting (see Remark 2).

Regarding optimal control for problems with distributed-order fractional operators, the results are rare and reduce to the following two papers: [17,18]. Both works develop numerical methods while, in contrast, here we are interested in analytical results (not in numerical approaches). Moreover, our results are new and bring new insights. Indeed, in [17], the problem is considered with Riemann–Liouville distributed derivatives, while in our case we consider optimal control problems with Caputo distributed derivatives. We must also note an inconsistency in [17]: when one defines the control system with a Riemann–Liouville derivative, then in the adjoint system it should appear as a Caputo derivative—when one considers optimal control problems with a control system with Caputo derivatives, the adjoint equation should involve a Riemann–Liouville operator—as a consequence of integration by parts (cf. Lemma 1). This inconsistency has been corrected in [18], where optimal control problems with Caputo distributed derivatives (as in this paper) are considered. Unfortunately, there is still an inconsistency in the necessary optimality conditions of both [17,18]: the transversality conditions are written there exactly as in the classical case, with the multiplier vanishing at the end of the interval, while the correct condition, as we prove in our Theorem 1, should involve a distributed integral operator—see condition (3).

The text is organized as follows. We begin by recalling definitions and necessary results of the literature in Section 2 of preliminaries. Our original results are then given in Section 3. More precisely, we consider fractional optimal control problems where the dynamical system constraints depend on distributed-order fractional derivatives. We prove a weak version of Pontryagin's maximum principle for the considered distributed-order fractional problems (see Theorem 1) and investigate a Mangasarian-type sufficient optimality condition (see Theorem 2). An example, illustrating the usefulness of the obtained results, is given (see Examples 1 and 2). We end with Section 4 of conclusions, mentioning also some possibilities of future research.

2. Preliminaries

In this section, we recall necessary results and fix notations. We assume the reader to be familiar with the standard Riemann–Liouville and Caputo fractional calculi [19,20].

Let α be a real number in $[0,1]$ and let ψ be a non-negative continuous function defined on $[0,1]$ such that

$$\int_0^1 \psi(\alpha) d\alpha > 0.$$

This function ψ will act as a distribution of the order of differentiation.

Definition 1 (See [1]). *The left and right-sided Riemann–Liouville distributed-order fractional derivatives of a function $x : [a,b] \to \mathbb{R}$ are defined, respectively, by*

$$\mathbb{D}_{a+}^{\psi(\cdot)} x(t) = \int_0^1 \psi(\alpha) \cdot D_{a+}^\alpha x(t) d\alpha \quad \text{and} \quad \mathbb{D}_{b-}^{\psi(\cdot)} x(t) = \int_0^1 \psi(\alpha) \cdot D_{b-}^\alpha x(t) d\alpha,$$

where D_{a+}^α and D_{b-}^α are, respectively, the left and right-sided Riemann–Liouville fractional derivatives of order α.

Definition 2 (See [1]). *The left and right-sided Caputo distributed-order fractional derivatives of a function $x : [a,b] \to \mathbb{R}$ are defined, respectively, by*

$$^C\mathbb{D}_{a+}^{\psi(\cdot)} x(t) = \int_0^1 \psi(\alpha) \cdot {}^C D_{a+}^\alpha x(t) d\alpha \quad \text{and} \quad {}^C\mathbb{D}_{b-}^{\psi(\cdot)} x(t) = \int_0^1 \psi(\alpha) \cdot {}^C D_{b-}^\alpha x(t) d\alpha,$$

where $^C D_{a+}^\alpha$ and $^C D_{b-}^\alpha$ are, respectively, the left and right-sided Caputo fractional derivatives of order α.

As noted in [16], there is a relation between the Riemann–Liouville and the Caputo distributed-order fractional derivatives:

$$^C\mathbb{D}_{a+}^{\psi(\cdot)} x(t) = \mathbb{D}_{a+}^{\psi(\cdot)} x(t) - x(a) \int_0^1 \frac{\psi(\alpha)}{\Gamma(1-\alpha)} (t-a)^{-\alpha} d\alpha$$

and

$$^C\mathbb{D}_{b-}^{\psi(\cdot)} x(t) = \mathbb{D}_{b-}^{\psi(\cdot)} x(t) - x(b) \int_0^1 \frac{\psi(\alpha)}{\Gamma(1-\alpha)} (b-t)^{-\alpha} d\alpha.$$

Along the text, we use the notation

$$\mathbb{I}_{b-}^{1-\psi(\cdot)} x(t) = \int_0^1 \psi(\alpha) \cdot I_{b-}^{1-\alpha} x(t) d\alpha,$$

where $I_{b-}^{1-\alpha}$ represents the right Riemann–Liouville fractional integral of order $1-\alpha$.

The next result has an essential role in the proofs of our main results; that is, in the proofs of Theorems 1 and 2.

Lemma 1 (Integration by parts formula [16]). *Let x be a continuous function and y a continuously differentiable function. Then,*

$$\int_a^b x(t) \cdot {}^C\mathbb{D}_{a+}^{\psi(\cdot)} y(t) dt = \left[y(t) \cdot \mathbb{I}_{b-}^{1-\psi(\cdot)} x(t) \right]_a^b + \int_a^b y(t) \cdot \mathbb{D}_{b-}^{\psi(\cdot)} x(t) dt.$$

Next, we recall the standard notion of concave function, which will be used in Section 3.3.

Definition 3 (See [21]). *A function $h : \mathbb{R}^n \to \mathbb{R}$ is concave if*

$$h(\beta \theta_1 + (1-\beta)\theta_2) \geq \beta h(\theta_1) + (1-\beta) h(\theta_2)$$

for all $\beta \in [0,1]$ and for all θ_1, θ_2 in $\mathbb{R}^n$.

Lemma 2 (See [21]). *Let $h : \mathbb{R}^n \to \mathbb{R}$ be a continuously differentiable function. Then h is a concave function if and only if it satisfies the so called gradient inequality:*

$$h(\theta_1) - h(\theta_2) \geq \nabla h(\theta_1)(\theta_1 - \theta_2)$$

for all $\theta_1, \theta_2 \in \mathbb{R}^n$.

Finally, we recall a fractional version of Gronwall's inequality, which will be useful to prove the continuity of solutions in Section 3.1.

Lemma 3 (See [22]). *Let α be a positive real number and let $a(\cdot)$, $b(\cdot)$, and $u(\cdot)$ be non-negative continuous functions on $[0, T]$ with $b(\cdot)$ monotonic increasing on $[0, T)$. If*

$$u(t) \leq a(t) + b(t) \int_0^t (t-s)^{\alpha-1} u(s) ds,$$

then

$$u(t) \leq a(t) + \int_0^t \left[\sum_{n=0}^\infty \frac{(b(t)\Gamma(\alpha))^n}{\Gamma(n\alpha)} (t-s)^{n\alpha-1} u(s) \right] ds$$

for all $t \in [0, T)$.

3. Main Results

The basic problem of optimal control we consider in this work, denoted by (BP), consists in finding a piecewise continuous control $u \in PC$ and the corresponding piecewise smooth state trajectory $x \in PC^1$ solution of the distributed-order non-local variational problem

$$J[x(\cdot), u(\cdot)] = \int_a^b L(t, x(t), u(t)) \, dt \longrightarrow \max,$$
$$^C\mathbb{D}_{a+}^{\psi(\cdot)} x(t) = f(t, x(t), u(t)), \quad t \in [a, b], \quad \text{(BP)}$$
$$x(\cdot) \in PC^1, \quad u(\cdot) \in PC,$$
$$x(a) = x_a,$$

where functions L and f, both defined on $[a, b] \times \mathbb{R} \times \mathbb{R}$, are assumed to be continuously differentiable in all their three arguments: $L \in C^1$, $f \in C^1$. Our main contribution is to prove necessary (Section 3.2) and sufficient (Section 3.3) optimality conditions.

3.1. Sensitivity Analysis

Before we can prove necessary optimality conditions to problem (BP), we need to establish continuity and differentiability results on the state solutions for any control perturbation (Lemmas 4 and 5), which are then used in Section 3.2. The proof of Lemma 4 makes use of the following mean value theorem for integration, that can be found in any textbook of calculus (see Lemma 1 of [23]): if $F: [0, 1] \to \mathbb{R}$ is a continuous function and ψ is an integrable function that does not change the sign on the interval, then there exists a number $\tilde{\alpha}$, such that

$$\int_0^1 \psi(\alpha) F(\alpha) \, d\alpha = F(\tilde{\alpha}) \int_0^1 \psi(\alpha) \, d\alpha.$$

Lemma 4 (Continuity of solutions). *Let u^ϵ be a control perturbation around the optimal control u^*, that is, for all $t \in [a, b]$, $u^\epsilon(t) = u^*(t) + \epsilon h(t)$, where $h(\cdot) \in PC$ is a variation and $\epsilon \in \mathbb{R}$. Denote by x^ϵ its corresponding state trajectory, solution of*

$$^C\mathbb{D}_{a+}^{\psi(\cdot)} x^\epsilon(t) = f(t, x^\epsilon(t), u^\epsilon(t)), \quad x^\epsilon(a) = x_a.$$

Then, we have that x^ϵ converges to the optimal state trajectory x^* when ϵ tends to zero.

Proof. Starting from the definition, we have, for all $t \in [a, b]$, that

$$\left| {}^C\mathbb{D}_{a+}^{\psi(\cdot)} x^\epsilon(t) - {}^C\mathbb{D}_{a+}^{\psi(\cdot)} x^*(t) \right| = |f(t, x^\epsilon(t), u^\epsilon(t)) - f(t, x^*(t), u^*(t))|.$$

Then, by linearity,

$$\left| {}^C\mathbb{D}_{a+}^{\psi(\cdot)} x^\epsilon(t) - {}^C\mathbb{D}_{a+}^{\psi(\cdot)} x^*(t) \right| = \left| {}^C\mathbb{D}_{a+}^{\psi(\cdot)} (x^\epsilon(t) - x^*(t)) \right| = |f(t, x^\epsilon(t), u^\epsilon(t)) - f(t, x^*(t), u^*(t))|$$

and it follows, by definition of the distributed operator, that

$$\left| \int_0^1 \psi(\alpha) {}^C D_{a+}^{\alpha} (x^\epsilon(t) - x^*(t)) \, d\alpha \right| = |f(t, x^\epsilon(t), u^\epsilon(t)) - f(t, x^*(t), u^*(t))|.$$

Now, using the mean value theorem for integration, and denoting $m := \int_0^1 \psi(\alpha) \, d\alpha$, we obtain that there exists an $\tilde{\alpha}$ such that

$$\left| {}^C D_{a+}^{\tilde{\alpha}} (x^\epsilon(t) - x^*(t)) \right| \leq \frac{|f(t, x^\epsilon(t), u^\epsilon(t)) - f(t, x^*(t), u^*(t))|}{m}.$$

Clearly, one has

$$^C D_{a+}^{\tilde{\alpha}}(x^\epsilon(t) - x^*(t)) \leq \left|^C D_{a+}^{\tilde{\alpha}}(x^\epsilon(t) - x^*(t))\right| \leq \frac{|f(t, x^\epsilon(t), u^\epsilon(t)) - f(t, x^*(t), u^*(t))|}{m},$$

which leads to

$$x^\epsilon(t) - x^*(t) \leq I_{a+}^{\tilde{\alpha}}\left[\frac{|f(t, x^\epsilon(t), u^\epsilon(t)) - f(t, x^*(t), u^*(t))|}{m}\right].$$

Moreover, because f is Lipschitz-continuous, we have

$$\left|f(t, x^\epsilon, u^\epsilon) - f(t, x^*, u^*)\right| \leq K_1\left|x^\epsilon - x^*\right| + K_2\left|u^\epsilon - u^*\right|.$$

By setting $K = \max\{K_1, K_2\}$, it follows that

$$\left|x^\epsilon(t) - x^*(t)\right| \leq \frac{K}{m} I_{a+}^{\tilde{\alpha}}\left(\left|x^\epsilon(t) - x^*(t)\right| + \left|\epsilon h(t)\right|\right)$$

$$= \frac{K}{m}\left[|\epsilon| I_{a+}^{\tilde{\alpha}}\left(|h(t)|\right) + I_{a+}^{\tilde{\alpha}}\left(\left|x^\epsilon(t) - x^*(t)\right|\right)\right]$$

$$= \frac{K}{m}\left[|\epsilon| I_{a+}^{\tilde{\alpha}}\left(|h(t)|\right) + \frac{1}{\Gamma(\tilde{\alpha})}\int_a^t (t-s)^{\tilde{\alpha}-1}\left|x^\epsilon(s) - x^*(s)\right| ds\right]$$

for all $t \in [a, b]$. Now, by applying Lemma 3 (the fractional Gronwall inequality), it follows that

$$\left|x^\epsilon(t) - x^*(t)\right| \leq \frac{K}{m}\left[|\epsilon| I_{a+}^{\tilde{\alpha}}\left(|h(t)|\right) + |\epsilon|\int_a^t \left(\sum_{i=0}^\infty \frac{1}{\Gamma(i\tilde{\alpha})}(t-s)^{i\tilde{\alpha}-1} I_{a+}^{\tilde{\alpha}}\left(|h(s)|\right)\right) ds\right]$$

$$= |\epsilon|\frac{K}{m}\left[I_{a+}^{\tilde{\alpha}}\left(|h(t)|\right) + \int_a^t \left(\sum_{i=1}^\infty \frac{1}{\Gamma(i\tilde{\alpha}+1)}(t-s)^{i\tilde{\alpha}} I_{a+}^{\tilde{\alpha}}\left(|h(s)|\right)\right) ds\right]$$

$$\leq |\epsilon|\frac{K}{m}\left[I_{a+}^{\tilde{\alpha}}\left(|h(t)|\right) + \int_a^t \left(\sum_{i=1}^\infty \frac{\delta^{i\tilde{\alpha}}}{\Gamma(i\tilde{\alpha}+1)} I_{a+}^{\tilde{\alpha}}\left(|h(s)|\right)\right) ds\right].$$

The series in the last inequality is a Mittag–Leffler function and thus convergent. Hence, by taking the limit when ϵ tends to zero, we obtain the desired result: $x^\epsilon \to x^*$ for all $t \in [a, b]$. □

Lemma 5 (Differentiation of the perturbed trajectory). *There exists a function η defined on $[a, b]$ such that*

$$x^\epsilon(t) = x^*(t) + \epsilon \eta(t) + o(\epsilon).$$

Proof. Since $f \in C^1$, we have that

$$f(t, x^\epsilon, u^\epsilon) = f(t, x^*, u^*) + (x^\epsilon - x^*)\frac{\partial f(t, x^*, u^*)}{\partial x} + (u^\epsilon - u^*)\frac{\partial f(t, x^*, u^*)}{\partial u} + o(|x^\epsilon - x^*|, |u^\epsilon - u^*|).$$

Observe that $u^\epsilon - u^* = \epsilon h(t)$ and $u^\epsilon \to u^*$ when $\epsilon \to 0$ and, by Lemma 4, we have $x^\epsilon \to x^*$ when $\epsilon \to 0$. Thus, the residue term can be expressed in terms of ϵ only, that is, the residue is $o(\epsilon)$. Therefore, we have

$$^C D_{a+}^{\psi(\cdot)} x^\epsilon(t) =\,^C D_{a+}^{\psi(\cdot)} x^*(t) + (x^\epsilon - x^*)\frac{\partial f(t, x^*, u^*)}{\partial x} + \epsilon h(t)\frac{\partial f(t, x^*, u^*)}{\partial u} + o(\epsilon),$$

which leads to

$$\lim_{\epsilon \to 0}\left[\frac{^C D_{a+}^{\psi(\cdot)}(x^\epsilon - x^*)}{\epsilon} - \frac{(x^\epsilon - x^*)}{\epsilon}\frac{\partial f(t, x^*, u^*)}{\partial x} - h(t)\frac{\partial f(t, x^*, u^*)}{\partial u}\right] = 0,$$

meaning that

$$^C\mathbb{D}_{a+}^{\psi(\cdot)}\left(\lim_{\epsilon \to 0} \frac{x^\epsilon - x^*}{\epsilon}\right) = \lim_{\epsilon \to 0} \frac{x^\epsilon - x^*}{\epsilon} \frac{\partial f(t, x^*, u^*)}{\partial x} + h(t) \frac{\partial f(t, x^*, u^*)}{\partial u}.$$

We want to prove the existence of the limit $\lim_{\epsilon \to 0} \frac{x^\epsilon - x^*}{\epsilon} =: \eta$, that is, to prove that $x^\epsilon(t) = x^*(t) + \epsilon \eta(t) + o(\epsilon)$. This is indeed the case, since η is solution of the distributed order fractional differential equation

$$\begin{cases} ^C\mathbb{D}_{a+}^{\psi(\cdot)} \eta(t) = \frac{\partial f(t, x^*, u^*)}{\partial x} \eta(t) + \frac{\partial f(t, x^*, u^*)}{\partial u} h(t), \\ \eta(a) = 0. \end{cases}$$

The intended result is proven. □

3.2. Pontryagin's Maximum Principle of Distributed-Order

The following result is a necessary condition of Pontryagin type [24] for the basic distributed-order non-local optimal control problem (BP).

Theorem 1 (Pontryagin Maximum Principle for (BP)). *If $(x^*(\cdot), u^*(\cdot))$ is an optimal pair for (BP), then there exists $\lambda \in PC^1$, called the adjoint function variable, such that the following conditions hold for all t in the interval $[a, b]$:*

- *The optimality condition*

$$\frac{\partial L}{\partial u}(t, x^*(t), u^*(t)) + \lambda(t) \frac{\partial f}{\partial u}(t, x^*(t), u^*(t)) = 0; \qquad (1)$$

- *The adjoint equation*

$$\mathbb{D}_{b-}^{\psi(\cdot)} \lambda(t) = \frac{\partial L}{\partial x}(t, x^*(t), u^*(t)) + \lambda(t) \frac{\partial f}{\partial x}(t, x^*(t), u^*(t)); \qquad (2)$$

- *The transversality condition*

$$\mathbb{I}_{b-}^{1-\psi(\cdot)} \lambda(b) = 0. \qquad (3)$$

Proof. Let $(x^*(\cdot), u^*(\cdot))$ be the solution to problem (BP), $h(\cdot) \in PC$ be a variation, and ϵ a real constant. Define $u^\epsilon(t) = u^*(t) + \epsilon h(t)$, so that $u^\epsilon \in PC$. Let x^ϵ be the state corresponding to the control u^*, that is, the state solution of

$$^C\mathbb{D}_{a+}^{\psi(\cdot)} x^\epsilon(t) = f(t, x^\epsilon(t), u^\epsilon(t)), \quad x^\epsilon(a) = x_a. \qquad (4)$$

Note that $u^\epsilon(t) \to u^*(t)$ for all $t \in [a, b]$ whenever $\epsilon \to 0$. Furthermore,

$$\left.\frac{\partial u^\epsilon(t)}{\partial \epsilon}\right|_{\epsilon=0} = h(t). \qquad (5)$$

Something similar is also true for x^ϵ. Because $f \in C^1$, it follows from Lemma 4 that, for each fixed t, $x^\epsilon(t) \to x^*(t)$ as $\epsilon \to 0$. Moreover, by Lemma 5, the derivative $\left.\frac{\partial x^\epsilon(t)}{\partial \epsilon}\right|_{\epsilon=0}$ exists for each t. The objective functional at (x^ϵ, u^ϵ) is

$$J[x^\epsilon, u^\epsilon] = \int_a^b L(t, x^\epsilon(t), u^\epsilon(t)) \, dt.$$

Next, we introduce the adjoint function λ. Let $\lambda(\cdot)$ be in PC^1, to be determined. By the integration by parts formula (see Lemma 1),

$$\int_a^b \lambda(t) \cdot {}^C D_{a+}^{\psi(\cdot)} x^\epsilon(t) dt = \left[x^\epsilon(t) \cdot \mathbb{I}_{b-}^{1-\psi(\cdot)} \lambda(t) \right]_a^b + \int_a^b x^\epsilon(t) \cdot \mathbb{D}_{b-}^{\psi(\cdot)} \lambda(t) dt,$$

and one has

$$\int_a^b \lambda(t) \cdot {}^C D_{a+}^{\psi(\cdot)} x^\epsilon(t) dt - \int_a^b x^\epsilon(t) \cdot \mathbb{D}_{b-}^{\psi(\cdot)} \lambda(t) dt - x^\epsilon(b) \cdot \mathbb{I}_{b-}^{1-\psi(\cdot)} \lambda(b) + x^\epsilon(a) \cdot \mathbb{I}_{b-}^{1-\psi(\cdot)} \lambda(a) = 0.$$

Adding this zero to the expression $J[x^\epsilon, u^\epsilon]$ gives

$$\phi(\epsilon) = J[x^\epsilon, u^\epsilon] = \int_a^b \left[L(t, x^\epsilon(t), u^\epsilon(t)) + \lambda(t) \cdot {}^C D_{a+}^{\psi(\cdot)} x^\epsilon(t) - x^\epsilon(t) \cdot \mathbb{D}_{b-}^{\psi(\cdot)} \lambda(t) \right] dt$$
$$- x^\epsilon(b) \cdot \mathbb{I}_{b-}^{1-\psi(\cdot)} \lambda(b) + x^\epsilon(a) \cdot \mathbb{I}_{b-}^{1-\psi(\cdot)} \lambda(a),$$

which by (4) is equivalent to

$$\phi(\epsilon) = J[x^\epsilon, u^\epsilon] = \int_a^b \left[L(t, x^\epsilon(t), u^\epsilon(t)) + \lambda(t) \cdot f(t, x^\epsilon(t), u^\epsilon(t)) - x^\epsilon(t) \cdot \mathbb{D}_{b-}^{\psi(\cdot)} \lambda(t) \right] dt$$
$$- x^\epsilon(b) \cdot \mathbb{I}_{b-}^{1-\psi(\cdot)} \lambda(b) + x_a \cdot \mathbb{I}_{b-}^{1-\psi(\cdot)} \lambda(a).$$

Since the process $(x^*, u^*) = (x^0, u^0)$ is assumed to be a maximizer of problem (BP), the derivative of $\phi(\epsilon)$ with respect to ϵ must vanish at $\epsilon = 0$; that is,

$$0 = \phi'(0) = \frac{d}{d\epsilon} J[x^\epsilon, u^\epsilon]|_{\epsilon=0}$$
$$= \int_a^b \left[\frac{\partial L}{\partial x} \frac{\partial x^\epsilon(t)}{\partial \epsilon} \bigg|_{\epsilon=0} + \frac{\partial L}{\partial u} \frac{\partial u^\epsilon(t)}{\partial \epsilon} \bigg|_{\epsilon=0} + \lambda(t) \left(\frac{\partial f}{\partial x} \frac{\partial x^\epsilon(t)}{\partial \epsilon} \bigg|_{\epsilon=0} + \frac{\partial f}{\partial u} \frac{\partial u^\epsilon(t)}{\partial \epsilon} \bigg|_{\epsilon=0} \right) \right.$$
$$\left. - \mathbb{D}_{b-}^{\psi(\cdot)} \lambda(t) \frac{\partial x^\epsilon(t)}{\partial \epsilon} \bigg|_{\epsilon=0} \right] dt - \frac{\partial x^\epsilon(b)}{\partial \epsilon} \bigg|_{\epsilon=0} \mathbb{I}_{b-}^{1-\psi(\cdot)} \lambda(b),$$

where the partial derivatives of L and f, with respect to x and u, are evaluated at $(t, x^*(t), u^*(t))$. Rearranging the term and using (5), we obtain that

$$\int_a^b \left[\left(\frac{\partial L}{\partial x} + \lambda(t) \frac{\partial f}{\partial x} - \mathbb{D}_{b-}^{\psi(\cdot)} \lambda(t) \right) \frac{\partial x^\epsilon(t)}{\partial \epsilon} \bigg|_{\epsilon=0} + \left(\frac{\partial L}{\partial u} + \lambda(t) \frac{\partial f}{\partial u} \right) h(t) \right] dt - \frac{\partial x^\epsilon(b)}{\partial \epsilon} \bigg|_{\epsilon=0} \mathbb{I}_{b-}^{1-\psi(\cdot)} \lambda(b) = 0.$$

Setting $H(t, x, u, \lambda) = L(t, x, u) + \lambda f(t, x, u)$, it follows that

$$\int_a^b \left[\left(\frac{\partial H}{\partial x} - \mathbb{D}_{b-}^{\psi(\cdot)} \lambda(t) \right) \frac{\partial x^\epsilon(t)}{\partial \epsilon} \bigg|_{\epsilon=0} + \frac{\partial H}{\partial u} h(t) \right] dt - \frac{\partial x^\epsilon(b)}{\partial \epsilon} \bigg|_{\epsilon=0} \mathbb{I}_{b-}^{1-\psi(\cdot)} \lambda(b) = 0,$$

where the partial derivatives of H are evaluated at $(t, x^*(t), u^*(t), \lambda(t))$. Now, choosing

$$\mathbb{D}_{b-}^{\psi(\cdot)} \lambda(t) = \frac{\partial H}{\partial x}(t, x^*(t), u^*(t), \lambda(t)), \quad \text{with } \mathbb{I}_{b-}^{1-\psi(\cdot)} \lambda(b) = 0,$$

that is, given the adjoint equation (2) and the transversality condition (3), it yields

$$\int_a^b \frac{\partial H}{\partial u}(t, x^*(t), u^*(t), \lambda(t)) h(t) = 0$$

and, by the fundamental lemma of the calculus of variations [25], we have the optimality condition (1):

$$\frac{\partial H}{\partial u}(t, x^*(t), u^*(t), \lambda(t)) = 0.$$

This concludes the proof. □

Remark 1. *If we change the basic optimal control problem* (BP) *by changing the boundary condition given on the state variable at initial time, $x(a) = x_a$, to a terminal condition, then the optimality condition and the adjoint equation of the Pontryagin Maximum Principle (Theorem 1) remain exactly the same. Changes appear only on the transversality condition:*

- *A boundary condition at final/terminal time—that is, fixing the value $x(b) = x_b$ with $x(a)$ remaining free, leads to*
$$\mathbb{I}_{a-}^{1-\psi(\cdot)}\lambda(a) = 0;$$

- *In the case when no boundary conditions is given (i.e., both $x(a)$ and $x(b)$ are free), then we have*
$$\mathbb{I}_{b-}^{1-\psi(\cdot)}\lambda(b) = 0 \quad \text{and} \quad \mathbb{I}_{a-}^{1-\psi(\cdot)}\lambda(a) = 0.$$

Remark 2. *If $f(t, x, u) = u$, that is, ${}^C\mathbb{D}_{a+}^{\psi(\cdot)}x(t) = u(t)$, then our problem* (BP) *gives a basic problem of the calculus of variations, in the distributed-order fractional sense of [16]. In this very particular case, we obtain from our Theorem 1 the Euler–Lagrange equation of [16] (cf. Theorem 2 of [16]).*

Remark 3. *Our distributed-order fractional optimal control problem* (BP) *can be easily extended to the vector setting. Precisely, let $x := (x_1, \ldots, x_n)$ and $u := (u_1, \ldots, u_m)$ with $(n, m) \in \mathbb{N}^2$, such that $m \leq n$, and functions $f : [a, b] \times \mathbb{R}^n \times \mathbb{R}^m \to \mathbb{R}^n$ and $L : [a, b] \times \mathbb{R}^n \times \mathbb{R}^m \to \mathbb{R}$ be continuously differentiable with respect to all its components. If (x^*, u^*) is an optimal pair, then the following conditions hold for $t \in [a, b]$:*

- *The optimality conditions*
$$\frac{\partial L}{\partial u_i}(t, x^*(t), u^*(t)) + \lambda(t) \cdot \frac{\partial f}{\partial u_i}(t, x^*(t), u^*(t)) = 0, \quad i = 1, \ldots, m;$$

- *The adjoint equations*
$$\mathbb{D}_{b-}^{\psi(\cdot)}\lambda_j(t) = \frac{\partial L}{\partial x_j}(t, x^*(t), u^*(t)) + \lambda(t) \cdot \frac{\partial f}{\partial x_j}(t, x^*(t), u^*(t)), \quad j = 1, \ldots, n;$$

- *The transversality conditions*
$$\mathbb{I}_{b-}^{1-\psi(\cdot)}\lambda_j(b) = 0, \quad j = 1, \ldots, n. \tag{6}$$

Definition 4. *The candidates to solutions of* (BP)*, obtained by the application of our Theorem 1, will be called* (Pontryagin) *extremals.*

We now illustrate the usefulness of our Theorem 1 with an example.

Example 1. *The triple $(\tilde{x}, \tilde{u}, \lambda)$ given by $\tilde{x}(t) = t^2$, $\tilde{u}(t) = \dfrac{t(t-1)}{\ln t}$, and $\lambda(t) = 0$, for $t \in [0, 1]$, is an extremal of the following distributed-order fractional optimal control problem:*

$$J[x(\cdot), u(\cdot)] = \int_0^1 -\left(x(t) - t^2\right)^2 - \left(u - \frac{t(t-1)}{\ln t}\right)^2 \longrightarrow \max,$$
$${}^C\mathbb{D}_{0+}^{\psi(\cdot)}x(t) = u(t), \quad t \in [0, 1], \tag{7}$$
$$x(0) = 0.$$

Indeed, by defining the Hamiltonian function as

$$H(t, x, u, \lambda) = - \left[(x - t^2) + \left(u - \frac{t(t-1)}{\ln t} \right)^2 \right] + \lambda u, \quad (8)$$

it follows:

- From the optimality condition $\frac{\partial H}{\partial u} = 0$,

$$\lambda(t) = 2 \left(u - \frac{t(t-1)}{\ln t} \right); \quad (9)$$

- From the adjoint equation $\mathbb{D}_{0+}^{\psi(\alpha)} \lambda(t) = \frac{\partial H}{\partial x}$,

$$\mathbb{D}_{0+}^{\psi(\alpha)} \lambda(t) = -2(x - t^2); \quad (10)$$

- From the transversality condition,

$$\mathbb{I}_{b-}^{1-\psi(\alpha)} \lambda(b) = 0. \quad (11)$$

We easily see that (9), (10) and (11) are satisfied for

$$x(t) = t^2, \quad u(t) = \frac{t(t-1)}{\ln t}, \quad \lambda(t) = 0.$$

3.3. Sufficient Condition for Global Optimality

We now prove a Mangasarian type theorem for the distributed-order fractional optimal control problem (BP).

Theorem 2. *Consider the basic distributed-order fractional optimal control problem* (BP). *If* $(x, u) \to L(t, x, u)$ *and* $(x, u) \to f(t, x, u)$ *are concave and* $(\tilde{x}, \tilde{u}, \lambda)$ *is a Pontryagin extremal with* $\lambda(t) \geq 0$, $t \in [a, b]$, *then*

$$J[\tilde{x}, \tilde{u}] \geq J[x, u]$$

for any admissible pair (x, u).

Proof. Because L is concave as a function of x and u, we have from Lemma 2 that

$$L(t, \tilde{x}(t), \tilde{u}(t)) - L(t, x(t), u(t)) \geq \frac{\partial L}{\partial x}(t, \tilde{x}(t), \tilde{u}(t)) \cdot (\tilde{x}(t) - x(t)) + \frac{\partial L}{\partial u}(t, \tilde{x}(t), \tilde{u}(t)) \cdot (\tilde{u}(t) - u(t))$$

for any control u and its associated trajectory x. This gives

$$J[\tilde{x}(\cdot), \tilde{u}(\cdot)] - J[x(\cdot), u(\cdot)] = \int_a^b [L(t, \tilde{x}(t), \tilde{u}(t)) - L(t, x(t), u(t))] dt$$

$$\geq \int_a^b \left[\frac{\partial L}{\partial x}(t, \tilde{x}(t), \tilde{u}(t)) \cdot (\tilde{x}(t) - x(t)) + \frac{\partial L}{\partial u}(t, \tilde{x}(t), \tilde{u}(t)) \cdot (\tilde{u}(t) - u(t)) \right] dt \quad (12)$$

$$= \int_a^b \left[\frac{\partial L}{\partial x}(t, \tilde{x}(t), \tilde{u}(t)) \cdot (\tilde{x}(t) - x(t)) - \frac{\partial L}{\partial u}(t, \tilde{x}(t), \tilde{u}(t)) \cdot (\tilde{u}(t) - u(t)) \right] dt.$$

From the adjoint equation (2), we have

$$\frac{\partial L}{\partial x}(t, \tilde{x}(t), \tilde{u}(t)) = \mathbb{D}_{b-}^{\psi(\cdot)} \lambda(t) - \lambda(t) \frac{\partial f}{\partial x}(t, \tilde{x}(t), \tilde{u}(t)).$$

From the optimality condition (1), we know that

$$\frac{\partial L}{\partial u}(t, \tilde{x}(t), \tilde{u}(t)) = -\lambda(t)\frac{\partial f}{\partial u}(t, \tilde{x}(t), \tilde{u}(t)).$$

It follows from (12) that

$$J[\tilde{x}(\cdot), \tilde{u}(t)] - J[x(\cdot), u(\cdot)] \geq \int_a^b \left(\mathbb{D}_{b-}^{\psi(\cdot)}\lambda(t) - \lambda(t)\frac{\partial f}{\partial x}(t, \tilde{x}(t), \tilde{u}(t)) \right) \cdot (\tilde{x}(t) - x(t))$$
$$- \lambda(t)\frac{\partial f}{\partial u}(t, \tilde{x}(t), \tilde{u}(t)) \cdot (\tilde{u}(t) - u(t)) \, dt. \quad (13)$$

Using the integration by parts formula of Lemma 1,

$$\int_a^b \lambda(t) \cdot {}^C\mathbb{D}_{a+}^{\psi(\cdot)}(\tilde{x}(t) - x(t)) \, dt = \left[(\tilde{x}(t) - x(t)) \cdot \mathbb{I}_{b-}^{1-\psi(\cdot)}\lambda(t) \right]_a^b + \int_a^b (\tilde{x}(t) - x(t)) \cdot \mathbb{D}_{b-}^{\psi(\cdot)}\lambda(t) dt,$$

meaning that

$$\int_a^b (\tilde{x}(t) - x(t)) \cdot \mathbb{D}_{b-}^{\psi(\cdot)}\lambda(t) dt$$
$$= \int_a^b \lambda(t) \cdot {}^C\mathbb{D}_{a+}^{\psi(\cdot)}(\tilde{x}(t) - x(t)) \, dt - \left[(\tilde{x}(t) - x(t)) \cdot \mathbb{I}_{b-}^{1-\psi(\cdot)}\lambda(t) \right]_a^b. \quad (14)$$

Substituting (14) into (13), we get

$$J[\tilde{x}(\cdot), \tilde{u}(\cdot)] - J[x(\cdot), u(\cdot)] \geq \int_a^b \lambda(t) \left[f(t, \tilde{x}(t), \tilde{u}(t)) \right.$$
$$\left. - f(t, x(t), u(t)) - \frac{\partial f}{\partial x}(t, \tilde{x}(t), \tilde{u}(t)) \cdot (\tilde{x}(t) - x(t)) - \frac{\partial f}{\partial u}(t, \tilde{x}(t), \tilde{u}(t)) \cdot (\tilde{u}(t) - u(t)) \right] dt.$$

Finally, taking into account that $\lambda(t) \geq 0$ and f is concave in both x and u, we conclude that $J[\tilde{x}(\cdot), \tilde{u}(\cdot)] - J[x(\cdot), u(\cdot)] \geq 0$. □

Example 2. *The extremal $(\tilde{x}, \tilde{u}, \lambda)$ given in Example 1 is a global minimizer for problem (7). This is easily checked from Theorem 2 since the Hamiltonian defined in (8) is a concave function with respect to both variables x and u and, furthermore, $\lambda(t) \equiv 0$. In Figure 1, we give the plots of the optimal solution to problem (7).*

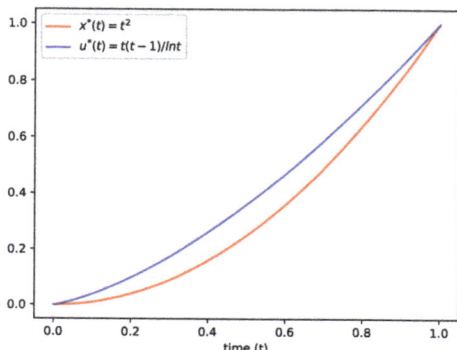

Figure 1. The optimal control u^* and corresponding optimal state variable x^*, solution of problem (7).

4. Conclusions

In this paper we investigated fractional optimal control problems depending on distributed-order fractional operators. We have proven a necessary optimality condition of Pontryagin's type and a Mangasarian-type sufficient optimality condition. The new results were illustrated with an example. As for future work, it would be interesting to develop proper numerical approaches to solve problems of optimal control with distributed-order fractional derivatives. In this direction, the approaches found in [17,18] can be easily adapted.

Author Contributions: The authors equally contributed to this paper, read and approved the final manuscript: Formal analysis, F.N. and D.F.M.T.; Investigation, F.N. and D.F.M.T.; Writing—original draft, F.N. and D.F.M.T.; Writing—review and editing, F.N. and D.F.M.T. All authors have read and agreed to the published version of the manuscript.

Funding: This research was funded by the Portuguese Foundation for Science and Technology (FCT), grant number UIDB/04106/2020 (CIDMA). Ndaïrou was also supported by FCT through the PhD fellowship PD/BD/150273/2019.

Acknowledgments: The authors are grateful to two anonymous reviewers for several comments and suggestions that have helped them to improve the manuscript.

Conflicts of Interest: The authors declare no conflict of interest.

References

1. Caputo, M. *Elasticità e Dissipazione*; Zanichelli: Bologna, Italy, 1969.
2. Caputo, M. Mean fractional-order-derivatives differential equations and filters. *Ann. Univ. Ferrara Sez. VII* **1995**, *41*, 73–84.
3. Bagley, R.L.; Torvik, P.J. On the existence of the order domain and the solution of distributed order equations. I. *Int. J. Appl. Math.* **2000**, *2*, 865–882.
4. Bagley, R.L.; Torvik, P.J. On the existence of the order domain and the solution of distributed order equations. II. *Int. J. Appl. Math.* **2000**, *2*, 965–987.
5. Caputo, M. Distributed order differential equations modelling dielectric induction and diffusion. *Fract. Calc. Appl. Anal.* **2001**, *4*, 421–442.
6. Diethelm, K.; Ford, N.J.; Freed, A.D.; Luchko, Y. Algorithms for the fractional calculus: A selection of numerical methods. *Comput. Methods Appl. Mech. Engrg.* **2005**, *194*, 743–773. [CrossRef]
7. Atanackovic, T.M. A generalized model for the uniaxial isothermal deformation of a viscoelastic body. *Acta Mech.* **2002**, *159*, 77–86. [CrossRef]
8. Lorenzo, C.F.; Hartley, T.T. Variable order and distributed order fractional operators. *Nonlinear Dynam.* **2002**, *29*, 57–98. [CrossRef]
9. Mainardi, F.; Mura, A.; Pagnini, G.; Gorenflo, R. Time-fractional diffusion of distributed order. *J. Vib. Control* **2008**, *14*, 1267–1290. [CrossRef]
10. Derakhshan, M.; Aminataei, A. Asymptotic Stability of Distributed-Order Nonlinear Time-Varying Systems with the Prabhakar Fractional Derivatives. *Abstr. Appl. Anal.* **2020**, *2020*, 1896563. [CrossRef]
11. Van Bockstal, K. Existence and uniqueness of a weak solution to a non-autonomous time-fractional diffusion equation (of distributed order). *Appl. Math. Lett.* **2020**, *109*, 106540. [CrossRef]
12. Zaky, M.A.; Machado, J.T. Multi-dimensional spectral tau methods for distributed-order fractional diffusion equations. *Comput. Math. Appl.* **2020**, *79*, 476–488. [CrossRef]
13. Ali, M.S.; Shamsi, M.; Khosravian-Arab, H.; Torres, D.F.M.; Bozorgnia, F. A space-time pseudospectral discretization method for solving diffusion optimal control problems with two-sided fractional derivatives. *J. Vib. Control* **2019**, *25*, 1080–1095. [CrossRef]
14. Nemati, S.; Lima, P.M.; Torres, D.F.M. A numerical approach for solving fractional optimal control problems using modified hat functions. *Commun. Nonlinear Sci. Numer. Simul.* **2019**, *78*, 104849. [CrossRef]
15. Sidi Ammi, M.R.; Torres, D.F.M. Optimal control of a nonlocal thermistor problem with ABC fractional time derivatives. *Comput. Math. Appl.* **2019**, *78*, 1507–1516. [CrossRef]
16. Almeida, R.; Morgado, M.L. The Euler-Lagrange and Legendre equations for functionals involving distributed-order fractional derivatives. *Appl. Math. Comput.* **2018**, *331*, 394–403. [CrossRef]

17. Zaky, M.A.; Machado, J.A.T. On the formulation and numerical simulation of distributed-order fractional optimal control problems. *Commun. Nonlinear Sci. Numer. Simul.* **2017**, *52*, 177–189. [CrossRef]
18. Zaky, M.A. A Legendre collocation method for distributed-order fractional optimal control problems. *Nonlinear Dyn.* **2018**, *91*, 2667–2681. [CrossRef]
19. Almeida, R.; Pooseh, S.; Torres, D.F.M. *Computational Methods in the Fractional Calculus of Variations*; Imperial College Press: London, UK, 2015. [CrossRef]
20. Samko, S.G.; Kilbas, A.A.; Marichev, O.I. *Fractional Integrals and Derivatives*; Gordon and Breach Science Publishers: Yverdon, Switzerland, 1993.
21. Rockafellar, R.T. *Convex Analysis*; Princeton University Press: Princeton, NJ, USA, 1970.
22. Ye, H.; Gao, J.; Ding, Y. A generalized Gronwall inequality and its application to a fractional differential equation. *J. Math. Anal. Appl.* **2007**, *328*, 1075–1081. [CrossRef]
23. Cao, L.; Li, Y.; Tian, G.; Liu, B.; Chen, Y. Time domain analysis of the fractional order weighted distributed parameter Maxwell model. *Comput. Math. Appl.* **2013**, *66*, 813–823. [CrossRef]
24. Pontryagin, L.S.; Boltyanskii, V.G.; Gamkrelidze, R.V.; Mishchenko, E.F. *The Mathematical Theory of Optimal Processes*; A Pergamon Press Book; The Macmillan Co.: New York, NY, USA, 1964.
25. Reid, W.T. Ramifications of the fundamental lemma of the calculus of variations. *Houst. J. Math.* **1978**, *4*, 249–262.

© 2020 by the authors. Licensee MDPI, Basel, Switzerland. This article is an open access article distributed under the terms and conditions of the Creative Commons Attribution (CC BY) license (http://creativecommons.org/licenses/by/4.0/).

MDPI
St. Alban-Anlage 66
4052 Basel
Switzerland
Tel. +41 61 683 77 34
Fax +41 61 302 89 18
www.mdpi.com

Axioms Editorial Office
E-mail: axioms@mdpi.com
www.mdpi.com/journal/axioms

www.ingramcontent.com/pod-product-compliance
Lightning Source LLC
LaVergne TN
LVHW070620100526
838202LV00012B/690